Sustainable Construction Supervisor

Trainee Guide
First Edition

Pearson

Harlow, England • London • New York • Boston • San Francisco • Toronto • Sydney
Dubai • Singapore • Hong Kong • Tokyo • Seoul • Taipei • New Delhi
Cape Town • São Paulo • Mexico City • Madrid • Amsterdam • Munich • Paris • Milan

National Center for Construction Education and Research

President: Don Whyte
Director of Product Development: Daniele Stacey
Sustainable Construction Supervisor Project Manager: Daniele Stacey
Production Manager: Tim Davis
Quality Assurance Coordinator: Debie Ness
Editor: Chris Wilson
Desktop Publishing Coordinator: James McKay
Production Specialist: Laura Wright
Cover Photo: Richard Berenholtz/The Durst Organization, Inc.

Writing and development services Virginia Tech/Myers-Lawson School of Construction
Dr. Annie Pearce
Dr. Christine Fiori

Editorial and production services provided by Topaz Publications, Liverpool, NY
Project Manager: Veronica Westfall
Desktop Publisher: Joanne Hart
Art Director: Megan Paye
Permissions Editors: Andrea LaBarge, Jackie Vidler

Pearson Education, Inc.
Product Manager: Lori Cowen
Operations Specialist: Deidra Skahill
Cover Design: Tim Davis

This book was set in Palatino and Univers and was printed and bound by LSC Communications. The cover was printed by LSC Communications.

The cover of this book was printed on Sustainable Forestry Initiative®(SFI)® approved paper LSC Communications. The text pages of this book were also printed and bound on approved paper by LSC Communications.

SUSTAINABLE FORESTRY INITIATIVE Certified Chain of Custody Promoting Sustainable Forestry www.sfiprogram.org

For more information regarding the Sustainable Forestry Initiative please visit **www.sfiprogram.org**.

nccer

ISBN 13: 978-0-13-215415-4

Pearson

Preface

To the Trainee

This training program and text have been developed for the Sustainable Construction Supervisor, an individual whose role it is to manage construction operations to support LEED certification. While the specific responsibilities of the Sustainable Construction Supervisor may vary from project to project and from company to company, the core functions remain the same: to ensure that sustainable building practices are put in place and that the LEED rating system meets the construction phase LEED credits targeted for the project.

This program is a collaborative effort of NCCER, the Green Building Certification Institute, the Myers-Lawson School of Construction at Virginia Tech, and a panel of Subject Matter Experts whose combined practical experience instills confidence in the credibility and usefulness of the training.

Critical issues affecting project sustainability include damage to the environment, use of resources, and generation of waste. Managing a project to increase its sustainability and success for LEED certification requires careful attention to construction operations, a key role played by the Sustainable Construction Supervisor.

Sustainable Construction Supervisor

If you are training under an NCCER Accredited Training Program Sponsor, successful completion of this program could yield nationally-recognized credentials, including a certificate and wallet card. Your training could also prepare you for certification as a Sustainable Construction Supervisor. To learn more, go to **www.nccer.org** or contact us at 1.888.622.3720.

The National Center for Construction Education and Research

We invite you to visit the NCCER website at **www.nccer.org** for information on the latest product releases and training, as well as online versions of the *Cornerstone* newsletter and Pearson's Contren® product catalog.

Your feedback is welcome. You may email your comments to **curriculum@nccer.org** or send general comments and inquiries to **info@nccer.org**.

Contren® Learning Series

The National Center for Construction Education and Research (NCCER) is a not-for-profit 501(c)(3) education foundation established in 1995 by the world's largest and most progressive construction companies and national construction associations. It was founded to address the severe workforce shortage facing the industry and to develop a standardized training process and curricula. Today, NCCER is supported by hundreds of leading construction and maintenance companies, manufacturers, and national associations. The Contren® Learning Series was developed by NCCER in partnership with Pearson Education, Inc., the world's largest educational publisher.

Some features of NCCER's Contren® Learning Series are as follows:

- An industry-proven record of success
- Curricula developed by the industry for the industry
- National standardization providing portability of learned job skills and educational credits
- Compliance with Office of Apprenticeship requirements for related classroom training (*CFR 29:29*)
- Well-illustrated, up-to-date, and practical information

NCCER also maintains a National Registry that provides transcripts, certificates, and wallet cards to individuals who have successfully completed modules of NCCER's Contren® Learning Series. *Training programs must be delivered by an NCCER Accredited Training Sponsor in order to receive these credentials.*

Introduction

Sustainable Construction Supervisor

The Green Building Certification Institute (GBCI) is pleased to endorse the Sustainable Construction Supervisor Training and Certification program. The construction industry is at the heart of green building, and the Sustainable Construction Supervisor is an example of a program that will equip construction professionals with the knowledge to implement green construction practices.

The demand for sustainable buildings is quickly transforming our built landscape. There are currently over 28,000 projects registered with GBCI that are seeking LEED® certification. These projects need construction professionals on the job who are armed with an in-depth understanding of green construction practices. The Sustainable Construction Supervisor program recognizes the vital role of construction professionals in the implementation of green building practices and design, and it delivers relevant training to the individuals who work on LEED and other green projects.

GBCI is proud to stand behind NCCER's leadership in the realm of green construction education as they raise the bar for the construction industry.

Beth Holst
Vice President, Credentialing
Green Building Certification Institute

GREEN BUILDING CERTIFICATION INSTITUTE

The Green Building Certification Institute (GBCI) is an independent, third-party organization committed to ensuring precision in the design, development, and implementation of the processes used to increase and measure green building performance (certification) and green building practice (credentialing). Established in 2008 to administer project certifications and professional credentials and certificates within the framework of the U.S. Green Building Council's Leadership in Energy and Environmental Design (LEED®) Green Building Rating Systems™, GBCI offers the marketplace the highest-level validation that LEED building certifications and LEED professional credentials have met specific, rigorous criteria.

U.S. GREEN BUILDING COUNCIL®

NCCER is a USGBC Education Provider. NCCER is committed to enhancing the ongoing professional development of building industry professionals and LEED Accredited Professionals (LEED AP) through appropriate Education Courses. As a USGBC Education Provider, NCCER has agreed to abide by USGBC-established operational and educational criteria and is subject to annual reviews and audits for quality assurance purposes. To learn more about the U.S. Green Building Council, the LEED rating system, or LEED, visit **www.usgbc.org**.

Major Contributors

Drs Annie Pearce and Christine Fiori

Major contributors to this module are Drs Annie Pearce and Christine Fiori of the Sustainable Facilities & Infrastructure Research Team of the Myers-Lawson School of Construction at Virginia Tech.

Dr. Annie Pearce is an Assistant Professor in the Myers-Lawson School of Construction at Virginia Tech specializing in sustainable facilities and infrastructure systems. Throughout her career, Annie has worked with practitioners in both public and private sectors to implement sustainability as part of building planning, design, construction, and operations. As a LEED Accredited Professional, Annie brings the latest in green building methods, technologies, and best practices to the classroom. Her specific areas of interest include metrics of sustainability for built facilities, green building materials and systems, cost modeling to support sustainability implementation, and in situ performance of sustainable facility technologies. Dr. Annie Pearce was the lead contributor to *Your Role in the Green Environment*, developed by the National Center for Construction Education and Research and published by Pearson.

Dr. Christine Fiori joined the Myers-Lawson School of Construction faculty at Virginia Polytechnic and State University in 2007 and serves as the Director for Undergraduate Programs. The focus of Dr. Fiori's research is in the area of the enhancement of construction education and the development of new and innovative pedagogical methods that embrace a more holistic approach to design and construction. These include the integration of leadership, sustainability and ancient technologies into course materials. She is the co-author of the McGraw-Hill textbook *Construction Management Fundamentals* and co-contributor to *Your Role in the Green Environment*, developed by the National Center for Construction Education and Research and published by Pearson. She has also developed a series of hands-on Green Construction Laboratories for NCCER that focus on the carpentry, electrical, and HVAC trades.

To ask questions or learn more about the ideas presented in this module, contact **sustainablefacilities@vt.edu**.

Foreword

The construction industry plays a key role in sustainability, as both provider and as custodian of the built environment, which is composed of industrial, residential, and non-residential facilities, as well as the civil infrastructure systems that provide the lifelines of any society. As documented extensively over the last few years, what the construction industry does, how it does it, and where and when, result in: (1) a significant yearly volume of economic activity; and (2) major direct and indirect impacts, such as (a) natural resource consumption, degradation, and depletion, (b) excessive energy and water consumption, (c) waste generation and accumulation, and (d) environmental impact and degradation.

The "green building" movement has emerged as a direct response to these impacts. This movement is supported by: (1) sustainability technical standards, such as the international sustainability standards for materials, products, systems, and services of the American Society for Testing and Materials – ASTM; (2) green building construction codes such as the new International Green Construction Code (IGCC) initiative of the International Code Council – ICC; (3) green building rating systems, such as the Leadership in Energy and Environmental Design (LEED) Rating System of the US Green Building Council – USGBC; and (4) green certification programs, such as the LEED Accredited Professionals certification program of the USGBC. These examples, among many more, highlight current efforts to influence the key stakeholders in any construction project, from owners, developers, and planners of construction projects, through designers and builders, to manufacturers and suppliers of construction materials, equipment, and products.

However, within this wide range of project stakeholders engaged in all aspects of green building, one key segment of this population, the construction supervisor, has not received much attention, until now. Construction supervisors are key to sustainable projects because they are the gatekeepers for all sustainability strategies determined during planning and design, and also, for the quality control for implementing those strategies during construction, commissioning, and turnover. Furthermore, construction supervisors have the knowledge necessary to ensure that smart decisions are made about what strategies to use, and they are becoming involved in projects earlier than ever before.

This book represents one more timely, tangible, and laudable effort of the National Center for Construction Education and Research (NCCER) to develop industry-driven standardized craft training programs. Working with a team of industry experts, Dr. Annie Pearce and Dr. Christine Fiori, from the Department of Building Construction of the Myers-Lawson School of Construction at the Virginia Polytechnic Institute and State University, have done an outstanding job in capturing the essence of sustainability for construction supervisors. While the book is primarily targeted to construction supervisors, in reality, the wealth, breadth, and depth of topics covered and of information provided should be of great interest and value for anyone interested in the practical implementation of sustainability in construction projects, particularly in the field. Furthermore, the efforts of Dr. Pearce and Dr. Fiori to bring together such a wide and varied array of information into this volume are impressive and commendable. This training program marks a major milestone on the road to more responsible construction practices, and provides unequivocal that the tipping point in green building has almost been reached, or may very close to doing so. It also will help prepare readers to take the Sustainable Construction Supervisor certification exam, which is being offered by NCCER, and endorsed by the Green Building Certification Institute.

In closing, NCCER should be recognized, in general, for its continuous commitment to its mission: to build a safe, productive, and sustainable workforce of craft professionals. However, more specifically and more importantly, NCCER needs to be commended for its foresight and its commitment to support the green building movement, and to the advancement of sustainability in the construction industry. I firmly believe that this book is clear evidence of this foresight and this commitment. In my opinion, it satisfies exemplarily the need in the construction industry to take sustainability directly to the field, and to the front line of project execution, and as a result, I heartily endorse it.

Dr. Jorge A. Vanegas,
Dean,
Sandy and Bryan Mitchell Master Builder
Endowed Chair,
Director, Center for Housing and Urban
Development (CHUD), and
Professor, Department of Architecture
College of Architecture, Texas A&M University
College Station, Texas

Acknowledgments

This curriculum was revised as a result of the farsightedness and leadership of these sponsor organizations. *Sustainable Construction Supervisor* would not exist were it not for the dedication and unselfish energy of those volunteers who served on the Authoring Team.

A sincere thanks is extended to them all.

Amanda Atkinson

LEED AP, BD + C, Senior Manager, Sustainable Services, Holder Construction Company

Kurt Langford

LEED AP, Project Executive, Suffolk Construction Company

Dave Conlon

LEED AP, Senior Project Manager, Hardin Construction

JJ McCarthy

LEED AP, Janus Development Services LLC

Howard Davis

Green Associate, Field Training Administrator, Brasfield and Gorrie

James G. Sullivan, Ph.D

LEED AP BD + C, Assistant Professor, M.E. Rinker, Sr. School of Building Construction, University of Florida

Boyd Worsham

LEED AP, Vice President of Construction Support, The Haskell Company

GBCI CMP

Green Building Certification Institute Credential Maintenance Program

LEED AP BD+C™ LEED AP O+M™ LEED AP ID+C™

Sustainable Construction Supervisor has been approved by the U.S. Green Building Council for 20 GBCI general *and* LEED-specific Continuing Education (CE) Hours for LEED Professionals. The LEED-specific categories for which this course have been approved are: LEED AP Building Design and Construction (BD+C); LEED AP Operations and Maintenance (O+M); and LEED AP Interior Design and Construction (ID+C).

The Green Building Certification Institute (GBCI) requires that LEED Green Associates earn 15 continuing education hours biennially (3 LEED-specific). LEED Accredited Professionals must earn 30 continuing education hours biennially (6 LEED-specific).

To report your CE hours with GBCI, go to GBCI's website at **www.gbci.org** and log into 'My Credentials' Account. A tutorial to walk you through the process of electronic reporting is available by going to the 'Earning and Reporting CE Hours' page at **www.gbci.org**.

You will need to report the CE hours per prescriptive category. NCCER recommends the following distribution of CE hours per prescriptive category for *Sustainable Construction Supervisor*:

No.	Category	Prescriptive Minumum Hours Required
I	Project Site Factors	5
II	Water Management	1
III	Project Systems and Energy Impacts	1
IV	Acquisition, Installation, and Management of Project Materials	5
V	Improvement to the Indoor Environment	4
VI	Stakeholder Involvement in Innovation	2
VII	Project Surrounding and Public Outreach	2
	Total Hours	20

Should you have any questions regarding reporting USGBC CE hours or the GBCI Credentialing Maintenance Program, please contact GBCI at 1-800-795-1746 or at **www.gbci.org**.

NCCER Partners

American Fire Sprinkler Association
Associated Builders and Contractors, Inc.
Associated General Contractors of America
Association for Career and Technical Education
Association for Skilled and Technical Sciences
Carolinas AGC, Inc.
Carolinas Electrical Contractors Association
Center for the Improvement of Construction Management and Processes
Construction Industry Institute
Construction Users Roundtable
Design Build Institute of America
Manufacturing Institute
Merit Contractors Association of Canada
Metal Building Manufacturers Association
NACE International
National Association of Minority Contractors
National Association of Women in Construction
National Insulation Association
National Ready Mixed Concrete Association
National Technical Honor Society
National Utility Contractors Association
NAWIC Education Foundation
North American Technician Excellence
Painting & Decorating Contractors of America
Portland Cement Association
SkillsUSA
Steel Erectors Association of America
U.S. Army Corps of Engineers
University of Florida
Women Construction Owners & Executives, USA

NCCER Business Partners

Contents

70201-10 Sustainable Construction Supervisor

Following the success and continued customer demand for green building training, NCCER and Virginia Tech have joined forces again to develop *Sustainable Construction Supervisor.* This new module was developed to instruct construction supervisors on sustainable construction management, the LEED rating system as it would apply to oversight of their projects and crews, and how to supervise and train their subcontractors and crews so that LEED points aren't unintentionally sacrificed. (20 Hours)

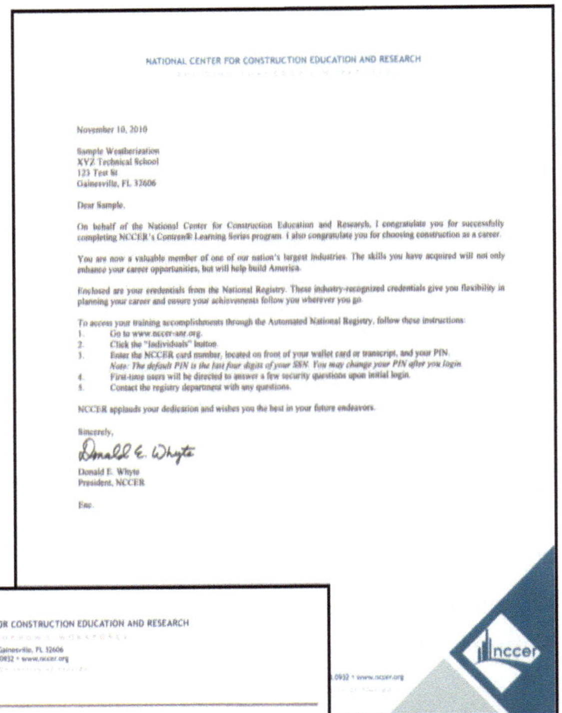
NATIONAL CENTER FOR CONSTRUCTION EDUCATION AND RESEARCH

November 10, 2010

Sample Weatherization
XYZ Technical School
123 Test St
Gainesville, FL 32606

Dear Sample,

On behalf of the National Center for Construction Education and Research, I congratulate you for successfully completing NCCER's Contren® Learning Series program. I also congratulate you for choosing construction as a career.

You are now a valuable member of one of our nation's largest industries. The skills you have acquired will not only enhance your career opportunities, but will help build America.

Enclosed are your credentials from the National Registry. These industry-recognized credentials give you flexibility in planning your career and ensure your achievements follow you wherever you go.

To access your training accomplishments through the Automated National Registry, follow these instructions:

1. Go to www.nccer-anr.org.
2. Click the "Individuals" button.
3. Enter the NCCER card number, located on front of your wallet card or transcript, and your PIN.
   Note: The default PIN is the last four digits of your SSN. You may change your PIN after you login.
4. First-time users will be directed to answer a few security questions upon initial login.
5. Contact the registry department with any questions.

NCCER applauds your dedication and wishes you the best in your future endeavors.

Sincerely,

Donald E. Whyte
President, NCCER

Enc.

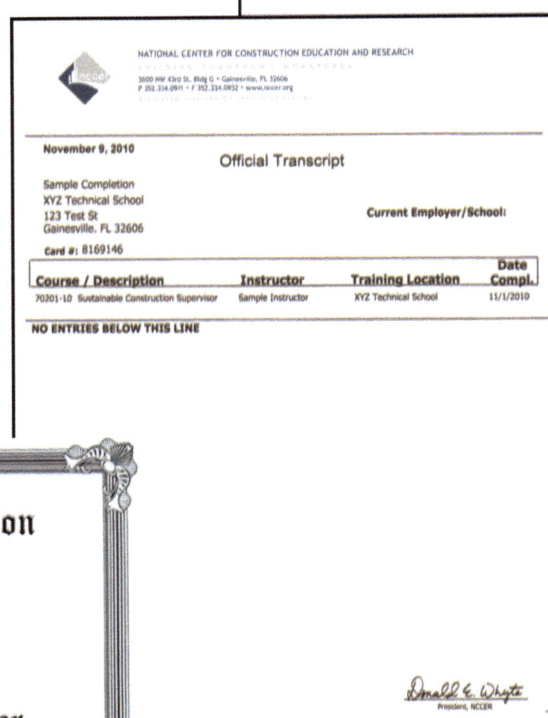
NATIONAL CENTER FOR CONSTRUCTION EDUCATION AND RESEARCH

3600 NW 43rd St, Bldg G • Gainesville, FL 32606
P 352.334.0911 • F 352.334.0932 • www.nccer.org

November 9, 2010

Official Transcript

Sample Completion
XYZ Technical School
123 Test St
Gainesville, FL 32606

Current Employer/School:

Card #: 8169146

Course / Description	Instructor	Training Location	Date Compl.
70201-10 Sustainable Construction Supervisor	Sample Instructor	XYZ Technical School	11/1/2010

NO ENTRIES BELOW THIS LINE

President, NCCER

National Center for Construction Education and Research

This is to certify that

Student 7 Sample

has fulfilled the requirements for

Sustainable Construction Supervisor

in NCCER's standardized training curriculum
this First day of December 2010

Endorsed By

Donald E. Whyte
President

Examples of credentials available on successful completion of Sustainable Construction Supervisor.

Contren® Curricula

NCCER's training programs comprise more than 80 construction, maintenance, pipeline, and utility areas and include skills assessments, safety training, and management education.

Boilermaking
Cabinetmaking
Carpentry
Concrete Finishing
Construction Craft Laborer
Construction Technology
Core Curriculum:
  Introductory Craft Skills
Drywall
Electrical
Electronic Systems Technician
Heating, Ventilating, and
  Air Conditioning
Heavy Equipment Operations
Highway/Heavy Construction
Hydroblasting
Industrial Coating and Lining
  Application Specialist
Industrial Maintenance
  Electrical and
  Instrumentation Technician
Industrial Maintenance
  Mechanic
Instrumentation
Insulating
Ironworking
Masonry
Millwright
Mobile Crane Operations
Painting
Painting, Industrial
Pipefitting
Pipelayer
Plumbing
Reinforcing Ironwork
Basic Rigger
Intermediate Rigger
Advanced Rigger
Scaffolding
Sheet Metal
Signal Person
Site Layout
Sprinkler Fitting
Tower Crane Operator
Welding

Green/Sustainable Construction

Building Auditor
Fundamentals of
  Weatherization
Introduction to Weatherization
Sustainable Construction
  Supervisor
Weatherization Crew Chief
Weatherization Technician
Your Role in the Green
  Environment

Energy

Introduction to the Power
  Industry
Introduction to Solar
  Photovoltaics
Introduction to Wind Energy
Power Industry Fundamentals
Power Generation Maintenance
  Electrician
Power Generation I&C
  Maintenance Technician
Power Generation Maintenance
  Mechanic

Pipeline

Control Center Operations,
  Liquid
Corrosion Control
Electrical and Instrumentation
Field Operations, Liquid
Field Operations, Gas
Maintenance
Mechanical

Safety

Field Safety
Safety Orientation
Safety Technology

Management

Fundamentals of Crew
  Leadership
Project Management
Project Supervision

Supplemental Titles

Applied Construction Math
Careers in Construction
Tools for Success

Spanish Translations

Rigging Fundamentals
  (Principios Básicos de
  Maniobras)
Carpentry Fundamentals
  (Introducción a la
  Carpintería, Nivel Uno)
Carpentry Forms
  (Formas para Carpintería,
  Nivel Trés)
Concete Finishing, Level One
  (Acabado de Concreto,
  Nivel Uno)
Core Curriculum:
  Introductory Craft Skills
  (Currículo Básico:
  Habilidades Introductorias
  del Oficio)
Drywall, Level One
  (Paneles de Yeso, Nivel Uno)
Electrical, Level One
  (Electricidad, Nivel Uno)
Field Safety
  (Seguridad de Campo)
Insulating, Level One
  (Aislamiento, Nivel Uno)
Masonry, Level One
  (Albañilería, Nivel Uno)
Pipefitting, Level One
  (Instalación de Tubería
  Industrial, Nivel Uno)
Reinforcing Ironwork, Level One
  (Herreria de Refuerzo,
  Nivel Uno)
Safety Orientation
  (Orientación de Seguridad)
Scaffolding
  (Andamios)
Sprinkler Fitting, Level One
  (Instalación de Rociadores,
  Nivel Uno)

Sustainable Construction Supervisor

70201-10

Trainees with successful module completions may be eligible for credentialing through NCCER's National Registry. To learn more, go to **www.nccer.org** or contact us at **1.888.622.3720**. Our website has information on the latest product releases and training, as well as online versions of our *Cornerstone* newsletter and Pearson's Contren® product catalog.

Your feedback is welcome. You may email your comments to **curriculum@nccer.org,** send general comments and inquiries to **info@nccer.org**, or use the User Update form at the back of this module.

V.1 12/10

70201-10

Sustainable Construction Supervisor

Objectives

When you have completed this module, you will be able to do the following:

1. Identify the short- and long-term impacts traditional construction practices have on the natural and human environment.
2. Describe the common problems encountered during the construction process that negatively impact sustainability factors for the project.
3. Propose ways to improve the sustainability of the project during the construction phase.
4. Develop different ways to achieve sustainability objectives when desired products or technologies are unavailable.
5. Describe the sustainability benefits of innovative actions that might earn a LEED innovation credit.
6. Use product information to evaluate whether products meet a given sustainability standard.
7. List and describe trade-offs for different products in terms of materials-related LEED credits.
8. Predict the impact of material substitutions on LEED credits and on budget, schedule, and other project requirements.
9. Review construction documents to analyze discrepancies with LEED requirements.
10. Verify that products, technologies, and project plans have required sustainability components and are being properly implemented.
11. Specify the roles and responsibilities of project stakeholders in achieving a green project.
12. Justify why timely and accurate tracking and verification of project documentation are critical to the project's success in LEED certification.

Prerequisites

Before you begin this module, it is recommended that you successfully complete *Your Role in the Green Environment*, Module 70101-09.

Trade Terms

Trade terms are highlighted in blue. Their definitions can be found in the Glossary at the back of this module.

Prerequisites

Before you begin this module, it is recommended that you successfully complete *Your Role in the Green Environment*, Module 70101-09.

Contents

Contents (continued)

Contents (continued)

Figures and Tables

Figures and Tables (continued)

Figures and Tables (continued)

Figures and Tables (continued)

"Participation in this curriculum and training has been personally rewarding. Through NCCER, we had the opportunity to share Best Practices across the industry to develop a program specific to the Sustainable Construction Supervisor, which provides an opportunity to display individual competency in this critical area."

Kurt Langford, LEED AP, Project Executive, Suffolk Construction Company

1.0.0 Sustainability and the Built Environment

The role of the Sustainable Construction Supervisor is critical to ensuring that green practices are employed in the built environment. The goal of this section is to describe why sustainability is important and illustrate how built facilities affect it. It also explains the links between actions on a construction site and the corresponding impacts and problems that affect the natural and human environment, both locally and globally. Specifically, this section describes the following:

- Impacts conventional construction practices have on the natural and human environment
- Approaches for **environmental stewardship** using green building **best practices**, technologies, materials, and methods
- Common challenges during construction that negatively impact project sustainability

1.1.0 Impacts of Land Development and Disturbance

As human beings, we develop land to meet our fundamental needs. Whether we build homes, stores, factories, or highways, our construction activities make a mark on the natural environment. Construction affects the land on which we build in multiple ways. How we arrange buildings into suburbs, towns, and cities is one important way that determines the impacts of buildings. Choosing a site for a new project contributes to the fabric of human development. It also predetermines the actions of building occupants in important ways. Finally, the ability of natural **ecosystems** to survive depends on how we arrange buildings on a project site. Survival also depends on what parts of the site we preserve.

1.1.1 Land Use, Development Patterns, and Implications for Site Selection

Where we decide to locate projects creates a legacy of impacts. Obvious impacts are generated as the site is cleared and prepared for construction when developing a project on a **greenfield** (previously undeveloped and undisturbed) site. However, the impacts of greenfield development go much further. Developing one parcel of land in a new area increases the likelihood that others will develop nearby. It also results in transportation impacts. Locating new development far from existing amenities requires occupants to drive when they need those amenities.

Conventional development also tends to segregate land into different uses. This can increase transportation requirements (*Figure 1*). Separating residential from commercial and public areas requires residents to drive to access those areas.

201F01.EPS

Figure 1 Urban sprawl leads to increased transportation requirements.

Did You Know?

There is an old saying in traffic engineering: "Capacity breeds congestion." For instance, that new lane on the crowded highway ultimately may make matters worse. Dynamic human systems behave in unpredictable ways. When a new traffic lane is brought online, the number of drivers choosing that route increases, thus returning congestion to earlier levels. This phenomenon also affects land use. For instance, businesses tend to locate near other businesses or near major transportation exits. This happens because they hope to build business from flow of traffic to other businesses. However, it can quickly cause significant human impacts in areas that are not well developed to handle it. When developing in new areas, be aware of the possible subsequent choices made by others.

Often the connections between contemporary developments, both residential and commercial, are limited. The occupants of a home on the edge of a housing development, such as the one in *Figure 1*, may have to drive several miles to reach a development on the other side of the property line. This is because of the branch-and-trunk road systems common in these types of developments.

1.1.2 Site Disturbance and Ecological Fragmentation

Choosing a project site has lasting impacts, both on the site to be developed and the sites that surround it. Human development on greenfield sites not only displaces whatever was on those sites before, but it also may encroach on valuable **agricultural land**. Sometimes it may threaten **wildlife habitat**. Even if no habitat is disturbed, wildlife will behave differently in the presence of human development. Disturbance also is likely to have an effect on surrounding water bodies, **wetlands**, and streams. Rain falling on paved or hard surfaces becomes **stormwater runoff** that flows quickly, in surges. It also tends to pick up heat from those surfaces, which disturbs the biology and chemistry of the body of water into which it flows.

Human developments may try to preserve open spaces for plants and animals. However, the overall effect is undeniable. *Figure 2* shows an example from the Amazonian rainforest. This satellite image shows that the actual area of human development is small. However, the effect is to divide larger areas of habitat into smaller pieces.

This reduces the amount of area for wildlife to occupy without encountering humans. It also leads to higher rates of extinction. This is because development separates species into smaller populations. These smaller populations are more vulnerable to disease and other threats.

201F02.EPS

Figure 2 Human development fragments ecosystems.

1.2.0 Impacts of Construction Materials

Not only does the choice of site for a project have impacts, but the materials used for the project also have a large **footprint**. By the time products arrive at the construction site, they have already consumed many resources and generated waste to get there. Installation of these materials in the project has direct and indirect effects. Finally, the nature of the materials determines the impacts they will have both as they are used and at the end of their **life cycle**.

1.2.1 Upstream Impacts

The life cycle of a construction material is divided into three phases: upstream of use, the use of the material itself, and downstream of use. **Upstream impacts** are all of the side effects a product has before it is actually used. Depending on the product, these effects may include the following:

- Harvesting the raw materials for the product from the natural environment
- Transporting those raw materials to factories for processing
- Processing the raw materials into finished components
- Transporting components to other factories for assembly
- Assembling the components into the finished product
- Packaging the finished product for transportation
- Transporting the finished product through the supply and distribution network to the project

Complex products, such as window and door assemblies, have many steps along the way. Simple products, such as aggregate, may have far fewer steps. Each of these steps uses materials and energy. Each step also produces waste. The total of all the energy consumed up until the point when the product reaches the project site is called embodied energy. Embodied energy is measured in units of energy per unit weight of a material.

Embodied energy varies greatly based on the type of material. *Figure 3* shows a comparison of embodied energy for different types of common construction materials. Synthetic products, such as carpet, have many different components. Each of these components has many manufacturing steps. These types of products have relatively high embodied energy per unit weight. Natural products, such as straw or aggregate, have lower embodied energy. This is because they are usually obtained close to the project site. They also do not have many manufacturing steps.

One interesting thing to note in *Figure 3* is the difference between virgin and **recycled materials**. Aluminum is the material with the highest embodied energy in the chart. It requires a large amount of energy to transform bauxite ore into alumina. It also requires more energy to convert alumina into aluminum. Using recycled aluminum eliminates this step. It can save up to 95 percent of the embodied energy. *Figure 3* also shows the difference between virgin and recycled steel.

Keeping track of all of the materials, energy, and waste associated with the product is known

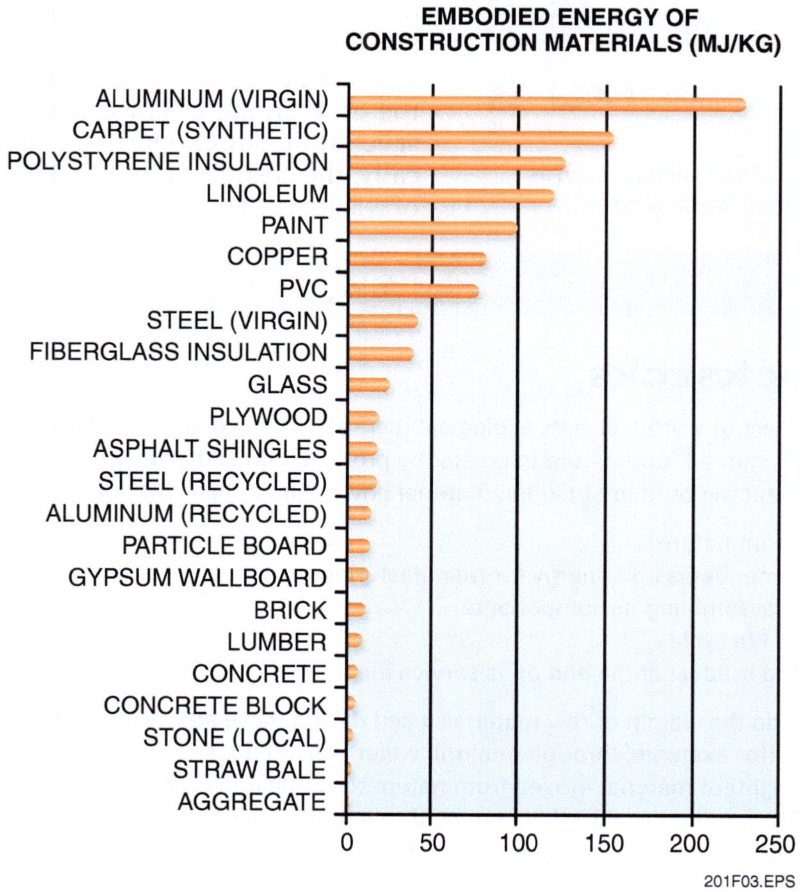

Figure 3 Construction materials have different levels of embodied energy (measured in MegaJoules per kilogram, or energy per unit weight).

as **life cycle assessment**. Focusing just on the upstream energy consumption is called embodied energy. Focusing just on the upstream materials (including the raw materials needed to produce the energy) is called a product's **ecological rucksack**.

In addition to the materials and energy consumed to make a product, waste is also produced. This waste is produced during harvesting of raw materials, manufacturing, and transport (*Figure 4*). A product's emissions, including **solid waste**, **liquid waste**, and air pollutants, are also tracked in a life cycle analysis.

201F04.EPS

Figure 4 The manufacture of building materials, such as steel, produces emissions of various types.

How and where the product is manufactured determine the amount of emissions it produces. These emissions are from the energy used in the manufacturing process. If it is complex with many components, the transportation required to bring those components together may be a large factor. Emissions, such as sedimentation from timber operations, may be tracked as soil loss. All of these factors are considered when evaluating the ecological history of a building material.

1.2.2 Impacts of Construction Means and Methods

How a product is installed affects the impacts. This is particularly true with respect to solid waste. Construction and demolition generate significant quantities of solid waste. The U.S. Environmental Protection Agency (EPA) has estimated this waste to be about 136 million tons per year. Standard new commercial construction generates between 2 and 25 pounds of solid waste per square foot. Many opportunities exist to reduce this impact.

On individual projects, design and construction decisions determine how much waste a project generates. A simple factor, such as orientation of tile on a floor, can influence the amount of product wasted by 5 percent or more (*Figure 5*). It also influences the labor cost of installation and the time. In the U.S., we often make choices that save labor but expend more materials. Historically, the cost of buying materials and disposing of waste was less expensive than the human labor

Did You Know?

Ecological Rucksacks

One way to think of the impact of a product is its ecological rucksack. The ecological rucksack of a product is the total weight of materials moved from nature to create the product, minus the weight of the product itself. Ecological rucksacks represent the burden of hidden material flows used to perform the following tasks:

1. Harvest raw materials from nature
2. Create and transport components and energy for manufacturing
3. Produce the product by assembling its components
4. Use the product over its life cycle
5. Recycle or dispose of the product at the end of its service life

Ecological rucksacks include the weight of raw materials used during these stages, soil moved during production of raw materials (for example, through mining), water consumed, and air consumed. The rucksack factor is the quantity (by weight) of material moved from nature to create one unit of the product. For example, the rucksack factor for virgin aluminum is 85:1. In contrast, the rucksack factor for recycled aluminum is 3.5:1, since there is no need to mine and process raw materials. The rucksack factor for a product can change over time as the raw materials become more scarce or the production process becomes more efficient.

Source: Friedrich Schmidt-Bleek, Wuppertal Institute for Climate, Environment and Energy. See www.gdrc.org for more information.

Think About It

Environmental Impact

There are several ways to express the environmental impact of a material. Embodied energy is measured as energy per unit weight. It refers to the energy it takes to make a product and get it ready for use. Ecological rucksack is measured as a ratio of the weight of raw materials versus finished materials. It refers to the amount of raw materials consumed to make a product. It includes the raw materials needed to make the energy to make the product. Which term includes more impacts?

201F05.EPS

Figure 5 The nature of a material and its installation determine waste ratios.

required to install. We have developed habits in construction that reflect this assumption. However, as raw materials become more expensive and waste disposal options are limited, the time has come to revisit these assumptions. Spending more time to reduce material waste may be a good investment. Especially during pre-construction planning, careful design of construction activities to reduce waste can be cost effective.

1.2.3 In-Use and Downstream Impacts

Building materials and products have impacts while they are in use in a building. They also have impacts at their **end of life cycle**. A life cycle assessment takes all of these phases into account.

One important consideration in evaluating a product is its durability and **fitness for purpose**. Choosing a product because its upstream environmental impacts are reduced may not be the best choice. If the product does not perform well and is replaced often, it is not effective. **Life cycle costing** is a method that considers a product's expected **service life**. Service life is the length of time a product is expected to remain in service before it fails or is otherwise unusable. Life cycle costing of a product takes into account the product's initial purchase price and cost of construction. It also considers the cost of operations, repairs, and replacement over the life cycle of the building. Products with longer service lives may cost more to purchase. However, they may have a lower life cycle cost when taking into account reduced future purchases.

Building products and materials also contribute to the quality of the indoor environment during their service lives. Some products, especially synthetic products, can have negative impacts on indoor air quality. The chemicals used in these products can offgas pollutants, such as **volatile organic compounds (VOCs)**, into the building (*Figure 6*). These products include adhesives, finishes, foams, and other materials. For this reason, a **low-emitting substance** should be selected whenever possible. Sometimes products must be used that will emit air pollutants during use. If this is the case, the building should be actively ventilated during installation. Furnishings and fixtures that will emit should be unpackaged and allowed to offgas away from the building if possible. Most **offgassing** happens within the first two weeks of a product's installation. This means that indoor air quality can be improved by including time in the schedule to ventilate the building prior to occupancy.

During operations and maintenance, building materials and products have differing requirements. Some products may require considerable maintenance using special cleaning products. Other products may require energy to operate and maintain. As mentioned earlier, choosing

201F06.EPS

Figure 6 Many common construction materials emit VOCs.

products with fewer upstream impacts is not always the best choice. For example, the finishes may require high-energy maintenance, such as regular stripping and waxing. Some products may cost more to purchase in the first place. However, their life cycle maintenance requirements may be much less (*Figure 7*). The whole life cycle of a product should be considered in making material selection decisions.

The end-of-life-cycle requirements of a product are also important in life cycle assessment. Some products present waste management problems. This may be due to the materials they contain. For example, products containing multiple types of plastic may be difficult to recycle and dangerous to incinerate. Many types of plastics generate dangerous chemical compounds when burned, such as dioxin. These compounds are released into the air. They can cause diseases, such as asthma and certain types of cancer. They can also fall out and be absorbed into the soil or groundwater, creating much more long-term pollution problems.

201F07.EPS

Figure 7 Choice of material determines ongoing maintenance.

Other products, such as structural elements of steel or aluminum, may have high-recycled value at the end of their life. Be sure to consider the following items as part of the life cycle cost assessment:

- The length of the product's service life
- The difficulty of disassembly and recycling
- The hazardous effects of disposal

Life cycle assessment can be useful in choosing products that are most sustainable for a building.

1.3.0 Impacts of Equipment Use and Construction Processes

The types of equipment used during construction impacts the natural environment. The processes used to install materials can also have impacts. Construction equipment creates air pollutant emissions by burning **fossil fuels**. Construction processes can also lead to significant damage to the soil on site. The utilities supplied temporarily to construction sites may be less efficient than permanent utilities. In addition, they may be damaged by equipment and not maintained as well as permanent utilities. Construction can also result in noise and **light pollution** that is disturbing to surrounding sites.

1.3.1 Emissions and Fossil Resource Use

Increased attention has focused on the impacts of construction equipment on air quality. Heavy construction equipment and other tools, such as portable generators, have significant impacts on local air quality (*Figure 8*). The burning of diesel fuel for heavy equipment in particular has been linked with human health problems, such as asthma and bronchitis. Diesel exhaust contains nitrogen oxides and extremely fine **particulates** that can lodge in human lungs. These can cause both short- and long-term health problems. Diesel exhaust has been linked to cancer in several studies. Children and adults with respiratory problems are especially vulnerable to health problems from diesel emissions.

The concerns about diesel exhaust led the EPA to create the Clean Construction USA program. Clean Construction USA provides incentives to reduce the emissions of diesel exhaust on construction sites. Tough new standards have been put into place for engines on new construction

201F08.EPS

Figure 8 Construction operations impact air quality.

equipment. However, existing equipment still causes problems. Diesel engines on construction equipment can be in service for as long as 25 to 30 years. Clean Construction USA encourages reducing idling time for heavy equipment. It also encourages installing after-market treatment technologies. These reduce emissions from heavy equipment.

1.3.2 Soil Compaction, Erosion, and Sedimentation

Construction projects often create significant damage to topsoil and subsoil (*Figure 9*). Plant life and the top layer of soil can be damaged by heavy equipment and excavation activities. Additionally, removing or damaging plant life reduces the stability of the soil. This causes it to erode more quickly. This **erosion** can damage both the construction site and neighboring sites, including nearby lakes and streams.

Topsoil contains organic matter and plant nutrients, along with soil microbes and insects. It supports plant life, regulates water flow, and helps control disease and pest outbreaks. Even if topsoil and subsoil are left in place on site, they can be compacted or damaged by heavy equipment or other site traffic. Parking vehicles on undisturbed grassy areas may appear to not cause damage, but the effects on the soil are real. Heat from vehicle engines and oil and fluid drippings from vehicles also cause damage to vegetation. These effects may not appear until after the project is complete. At that point, additional fertilizer, irrigation, and landscaping may be required to fix the damage. Fertilizers and pesticides also have negative impacts on local streams and waterways when they leave the site as stormwater runoff.

201F09.EPS

Figure 9 Heavy equipment damages plant life.

Airborne dust can also be a problem when soil is left exposed. This dust has impacts on both the natural environment and human health. Workers and neighbors can breathe in fine dust particles. This dust can cause or worsen problems, such as asthma, and can make breathing difficult. It also can travel long distances before settling, resulting in damage to the natural and manmade sites where it lands.

1.3.3 Temporary Utilities and Resource Consumption

Temporary utilities are used during construction operations before permanent utilities are installed. These include temporary power and water supplies. They may also include temporary lighting, heating, cooling, or ventilation for spaces in the building after it has been enclosed.

Temporary utilities are used to provide suitable working conditions for workers in the building. These utilities also maintain the necessary conditions for installation of interior materials. They may be used to perform the following functions:

- Manage moisture
- Cure concrete
- Install interior finishes
- Ventilate contaminants from an indoor space

Workers should be instructed regarding the proper, efficient, and safe use of these systems. Many temporary heating units are fired with combustible fuels. They must be carefully maintained to avoid dangerous working conditions and environmental damage. Workers should be sure to turn off temporary equipment when it is not needed. However, they must be instructed on when equipment should be left on to maintain environmental conditions in the building.

1.3.4 Noise and Light Pollution

Noise and light pollution are inevitable for some types of construction projects (*Figure 10*). Construction schedules may dictate that night construction is required. Alternatively, night construction may be scheduled to avoid congestion and increase safety. This is common in highway construction.

Light pollution occurs when excess illumination reflects off water droplets in the night sky. It also occurs when lights are directed outside the work area. It often represents an inefficient use of energy. It can disrupt both human and wildlife sleep cycles for neighboring sites. **Noise pollution** is also common for some types of construction activities. It occurs when unwanted vibrations leave the construction site and disturb neighboring sites. These vibrations may be either audible or inaudible, depending on the frequency. Noise pollution, if it is loud enough or long enough, can cause permanent hearing damage. However, even in smaller quantities it can cause sleep disruption, stress, and even raise blood pressure, especially for parties not involved in construction. Some areas have noise ordinances that must be observed during construction. Both construction schedules and equipment/method choices can be used to minimize the effects of light and noise pollution on the surrounding environment.

1.4.0 Social Impacts of Construction

In addition to the impacts of construction on the natural and human environment, it can also have an effect on human social and economic systems. The **corporate social responsibility (CSR)** of a company is becoming more important. It is defined by how its business affects people. This includes employees and people in the market. It also considers neighboring communities that are affected by the company's business practices.

201F10.EPS

Figure 10 Construction activities emit noise and light pollution.

What's wrong with this picture?

201SA01.EPS

What's wrong with this picture?

201SA02.EPS

What's wrong with this picture?

201SA03.EPS

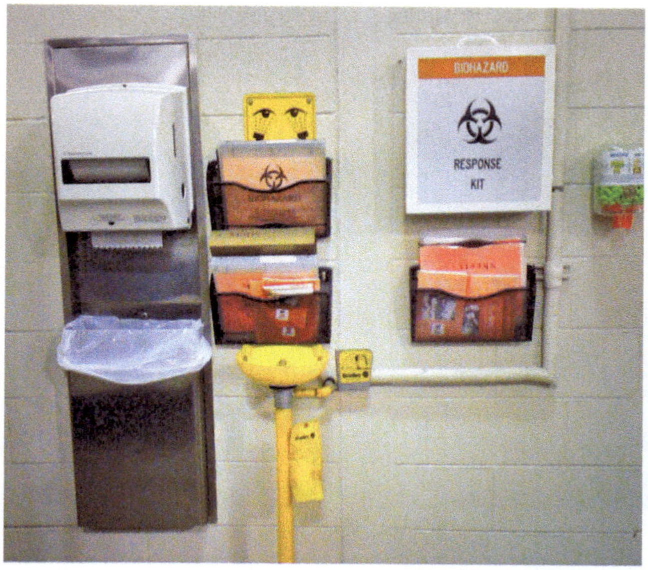

201F11.EPS

Figure 11 A safe and healthy workplace is part of social sustainability.

Increasingly, companies are measuring CSR as part of a **triple bottom line** approach to business. In addition to profit, companies measure the impacts of their business on society and on the natural environment. This broader set of criteria is important for corporate sustainability.

The social impacts of construction can be improved by creating a healthy workplace for construction workers. Social impacts can also be influenced by promoting diversity and equal opportunity and by using local resources to develop and support the project.

1.4.1 Healthy Workplace

With increased attention to construction safety, the job site has become a safer place to work. Many injuries have been prevented through better training, the proper use of personal protective equipment, and improved job practices.

Better industrial hygiene has also resulted in fewer long-term injuries and illnesses for the construction worker.

Complying with federal regulations for job safety is a must. For instance, in areas where workers are exposed to hazardous chemicals, proper wash areas and emergency response equipment should be present (*Figure 11*). Each job site must be evaluated for all known environmental risks before work begins.

The long-term health impacts of exposure to airborne chemicals and particulates are becoming better understood. Increasing ventilation of areas where these are present is important for workers during construction.

As we gain a better understanding of the health impacts of chemical exposure, the use of personal protective equipment becomes even more important. Green building projects seek to reduce the types of hazardous materials used. However, construction workers must still be aware of potential health risks and take measures to mitigate or avoid them.

1.4.2 Diversity and Equal Opportunity

A second key part of social sustainability is providing opportunities for workers to grow and develop in their profession. Diversity may be in terms of age, gender, ethnicity, or other factors. Diversity in the work force benefits construction companies by bringing together people with a variety of backgrounds and skills to complete a job (*Figure 12*). It can also provide a competitive advantage for companies by retaining employees.

Providing equal opportunity for employment, training, and advancement can be advantageous for construction companies. The costs of employee turnover are significant. Recruiting and training new employees can be expensive. The need to fill positions on a crew can mean that work is delayed or does not get done correctly. Providing equal opportunities for workers to advance helps in retaining existing employees. These workers bring additional experience, which helps contribute to a more sustainable project.

201F12.EPS

Figure 12 Diverse workplaces promote social sustainability.

1.4.3 Creating Positive Local Impacts

As the activity that builds the infrastructure of society, construction can create powerful local impacts. Construction firms can contribute to sustainability by building green projects. These projects can also employ resources that help to build the economic sustainability of the local community and region.

Hiring local businesses helps to support the local economy. It may also build capacity for future work. Using local products and materials also produces the same results. Just as local farmers' markets provide a way for citizens to obtain fresher and healthier food (*Figure 13*), construction projects that support local businesses improve the health of the local economy. They provide a way for the labor of local citizens to contribute to the buildings and facilities being built in their communities.

201F13.EPS

Figure 13 Just like a local farmer's market, using local businesses and products helps to support the local economy.

Using local products and services also improves the environment by reducing the need to transport goods and service providers from other places. The skills and capabilities built at the local level will contribute to more vibrant, resilient communities that can sustain themselves economically in the long term.

Review Questions

1. The embodied energy of a construction material is the amount of energy required to _____.
   a. extract the virgin material from the environment
   b. transport the product to the construction site
   c. recycle the product into new materials
   d. manufacture and transport all components of a product until the product reaches the site

2. Life cycle costing considers the product's _____.
   a. ease of installation
   b. service life
   c. recyclability
   d. salvage capability

3. End-of-life-cycle requirements are important to consider in LEED projects because some materials _____.
   a. offgas throughout their lifetime
   b. are not fit for their purpose
   c. require faster replacement
   d. require increased maintenance

4. In addition to requiring emissions standards for new equipment, Clean Construction USA encourages _____.
   a. the use of cleaner burning fuel
   b. the use of diesel fuel
   c. reduced idling time for heavy equipment
   d. reduced operating time for heavy equipment

5. Noise and light pollution from construction sites impact the _____.
   a. construction site
   b. construction schedule
   c. efficient use of manpower
   d. efficient use of energy

"The NCCER has been a great asset in the training of our Field Employees and this curriculum is one more step to creating the best Field Training program in the industry. With the emphasis of sustainability in our industry, it is imperative all of our employees receive training on this topic and this module only furthers that initiative."

Howard Davis, Green Associate, Field Training Administrator, Brasfield and Gorrie

2.0.0 Project Sustainability Goals: From Principles to Practice

There are many ways to improve the sustainability of a construction project. The goal of this section is to show the links between broad sustainability goals for a project and best practices on site to achieve them. This will help in the development of a project strategy to apply principles of environmental stewardship and comply with project sustainability goals. Specifically, this section describes how to perform the following tasks:

- Identify ways to improve the sustainability of the project during the construction phase of the project life cycle.
- Identify alternative means of achieving objectives when desired products or technologies are unavailable.
- Recognize sustainability implications of changes made during project implementation.
- Understand the Leadership in **Energy and Environmental Design (LEED)** rating system as a way to manage the construction process of a project.
- Choose courses of action that contribute to LEED certification.
- Describe the sustainability benefits associated with innovative actions that might earn a LEED **innovation credit**.

2.1.0 Balancing Impacts and Benefits of Construction

Construction projects influence the world both inside and beyond the site. As soon as site clearing begins, the impacts to the project site are easy to see. Impacts outside the project's boundary are not so easy to recognize. The materials and energy used during construction have a legacy of impacts. These impacts stem from the raw materials needed to make the product and the energy needed for manufacture and transportation.

2.1.1 Rules of Thumb for Managing Construction Impacts

The project sustainability team works to balance the benefits provided by a product with the impacts it creates. The first step in creating this balance is to identify the impacts caused by the project and the products and actions that cause them. Projects create negative impacts both inside and outside the site boundary (*Figure 14*).

Impacts inside the site boundary mainly involve disturbing or destroying natural ecosystems on the site. This also includes displacing animals, such as deer and rodents. Natural ecosystems provide a variety of functions that humans require to survive. These services include cleaning and purifying air and water, regenerating resources such as wood products, and stabilizing the Earth's climate.

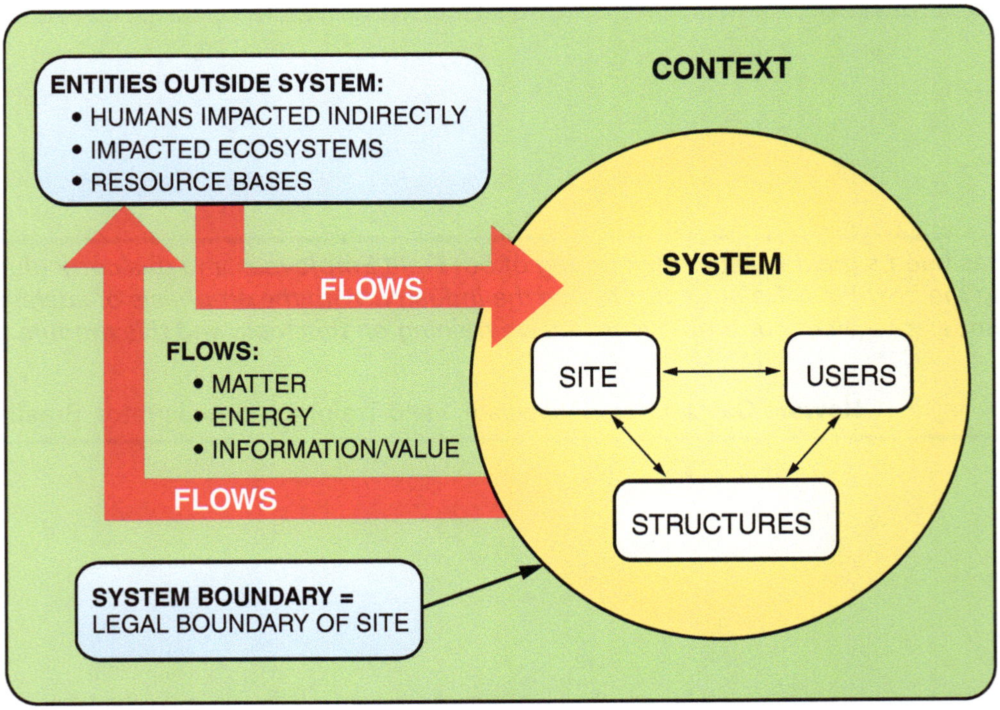

201F14.EPS

Figure 14 Construction projects are complex systems that affect entities both inside and outside the site boundary.

Did You Know?

Scientists have estimated the value of ecosystem services to be around 33 trillion dollars per year. This represents what it would cost to provide the fertile soil, fresh water, breathable air, and amenable climate that nature automatically provides.

Source: World Resources Institute (www.wri.org)

The project can also cause ecosystem disturbance across the site boundary. This disturbance may be due to pollutants, such as stormwater runoff, or vehicle emissions and noise. Choices we make about site layout, scheduling, staging, and construction means and methods affect the level of impact of the project on the site.

Impacts outside the site boundary are primarily caused by imports or exports of materials, water, energy, and waste across the system boundary. Most of the negative impacts are caused by transportation. However, these imports and exports also cause impacts at their sources or destinations. The selection of various products, services, vendors and suppliers, and construction methods influences the impacts of the project beyond the site boundary.

2.1.2 Regenerative Construction Practices

Regenerative construction practices go beyond minimum sustainability. They create or restore ecosystems. They also create resources that can be used by others. Regenerative construction takes its inspiration from natural processes. It creates capacity for human development both now and in the future. Building on ideas developed by architect Malcolm Wells, the Society of Building Science Educators has developed a checklist to evaluate a project's site and building(s). *Table 1* shows the factors that determine how regenerative a construction site is.

Table 1 Checklist of Regenerative Construction Factors – Site (after Wells 1969)

Destructive	Regenerative
Destroys pure air	Creates pure air
Destroys pure water	Creates pure water
Wastes rainwater	Stores rainwater
Produces no food	Produces its own food
Destroys rich soil	Creates rich soil
Wastes solar energy	Uses solar energy
Stores no solar energy	Stores solar energy
Destroys silence	Creates silence
Dumps its wastes unused	Consumes its own wastes
Needs cleaning and repair	Maintains itself
Disregards nature's cycles	Matches nature's cycles
Destroys wildlife habitat	Provides wildlife habitat
Destroys human habitat	Provides human habitat
Intensifies local weather	Moderates local weather
Unattractive	Beautiful

As shown in the checklist, regenerative site practices result in cleaner, better quality air, water, and soil than existed before the project. Regenerative project sites take advantage of and store natural resources, such as rainwater and solar energy. Mimicking natural processes, they consume their own wastes. They also turn waste into useful products. Some conversion of waste may occur on-site through composting and irrigation with wastewater. Other wastes are captured and recycled off-site. Regenerative sites create beautiful, quiet, attractive habitats for humans and wildlife.

Did You Know?

Carbon offsets, also known as green tags, can be purchased by project stakeholders to compensate for the effects of their activities on global climate change. Various companies offer these tags and use the money generated by them to undertake projects that sequester carbon or otherwise reduce greenhouse gases. For instance, they may use funds from green tags to plant trees that will absorb carbon, or develop clean energy sources to reduce the need for fossil fuel-based power. Other firms capture methane from landfills to prevent it from being vented to the atmosphere. Directories of carbon offset opportunities are available online (see www.carbonify.com, for example). Third-party certification of these activities is important to ensure that the purchase achieves the desired outcome.

They use **green space** to moderate the local climate, rather than allowing pavement, walls, and waste heat to intensify it (*Figure 15*). Finally, they are deliberately designed to be easy to maintain rather than creating unnecessary maintenance requirements.

Table 2 lists regenerative factors for buildings. Like regenerative sites, regenerative buildings use natural or passive systems. These are used instead of mechanical systems for lighting, heating, cooling, and circulation. These buildings are designed to require minimum maintenance. Some can even maintain themselves. They feature beautiful environments that are comfortable for humans. They are built of recycled materials and can themselves be recycled. They incorporate visible, appropriate responses to local environmental conditions and constraints and fit well within their context.

These rules of thumb for sustainable and regenerative projects provide a foundation for setting specific project sustainability goals. They also help in finding strategies to meet those goals. They provide general guidance on what actions lead to better project sustainability and what actions lead to problems. The next step is understanding how to use measurement to systematically increase project sustainability.

Table 2 Checklist of Regenerative Construction Factors – Buildings (Wells 1969)

Destructive	Regenerative
Excludes natural light	Uses natural light
Uses mechanical heating	Uses passive heating
Uses mechanical cooling	Uses passive cooling
Needs cleaning and repair	Maintains itself
Produces human discomfort	Provides human comfort
Uses fuel-powered circulation	Uses human-powered circulation
Pollutes indoor air	Purifies indoor air
Built of virgin materials	Built of recycled materials
Cannot be recycled	Can be recycled
Ignores local ecology	Respects local ecology
Bad neighbor	Good neighbor
Unattractive	Beautiful

201F15.EPS

Figure 15 Green roofs provide enclosure, stormwater management, and thermal control while cleaning the air and providing a habitat for birds and insects.

2.2.0 Managing Impacts Through Measurement

Management by objectives is an approach to projects that involves goal setting, and then measuring how well those goals are met. Approaching project sustainability in this way allows the Sustainable Construction Supervisor to monitor progress, identify potential problems, and celebrate success. The first step in this approach is to set project sustainability goals that will be used as a guide for the project.

2.2.1 Establishing Project Sustainability Goals and Objectives

There are many ways to set sustainability goals and objectives for a project. Some projects pursuing formal certification under a recognized rating system may choose to use the framework of the rating system as a starting point. Other projects may have goals outside of any particular framework. Goals are broad statements of what should be achieved for or by the project as a whole. Objectives are more specific and have the characteristics of being both measurable and time-bound. Both objectives and goals for a project should be attainable and relevant to the project and issues at hand. The SMART acronym provides a way to remember the characteristics of effective objectives for a project. These objectives should be **S**pecific, **M**easurable, **A**ttainable, **R**elevant, and **T**ime-bound.

Measurable means it is possible to definitively say whether or not an objective has been achieved. Some objectives can be expressed in terms of quantifiable performance—for instance, "reduce water use by 50 percent." These objectives require modeling or measurement to determine whether the specified level has been met. Others may be measurable in terms of yes or no; for instance, "use bio-based form release oil for all concrete forms."

Time-bound means that the point at which the objective will be evaluated is defined at the beginning. For instance, the objective to reduce water use by 50 percent does not explicitly specify the measurement period. Must this objective be met during construction? During a typical year of operation? The objective only makes sense if the time period for evaluation is specified. This objective must also be made more specific to be useful. The scope of the objective should be clear. What water uses will be counted? What is the baseline against which the reduction should be measured?

To be relevant, the objective should have a scope that is meaningful. It must be meaningful for the time period and project being considered. For instance, reducing water use may be meaningful for a building during operations, but is not relevant for a highway project during operations. However, water use might be relevant for certain aspects of the construction of a highway project, depending on the means and methods involved.

Finally, effective objectives should be attainable. They must be attainable given the resources and parameters of the project. Objectives are most useful when they require additional effort beyond common practice. However, they should be reasonably achievable when considering the additional costs and efforts of achieving them. Determining what makes an objective attainable is a balance. The balance is between the cost of meeting the objective and the benefits that can result. Setting unattainable objectives may result in unwise investment of resources. It may also be bad for morale when the unattainable objective is not met.

In general, sustainability goals for a project fall into three overlapping categories. These categories correspond to what is known as the triple bottom line (*Figure 16*). Goals and objectives should focus on the material rather than the method. For instance, specifying a goal of diverting 50 percent by weight of construction waste from landfill disposal retains flexibility for the project team to determine the most effective way to achieve it.

A Sustainable Construction Supervisor's role in the goal setting process will vary from project to project. In some cases, the Sustainable Construction Supervisor will be invited to share knowledge as an active member of the project sustainability team. Sustainability goals and objectives are typically established early in the project life cycle. This typically occurs during a brainstorming session that includes all key stakeholders. They may also be established by the owner at a general level before the project begins. These goals may be a part of the owner's project requirements—in this case, the project team will be involved in developing specific objectives to achieve the owner's goals. The team will also determine whether objectives are realistic based on what the project is and how it is to be constructed. The expertise of a Sustainable Construction Supervisor will be most

Think About It

How could the objective "Reduce water use by 50 percent" be rewritten to meet SMART criteria?

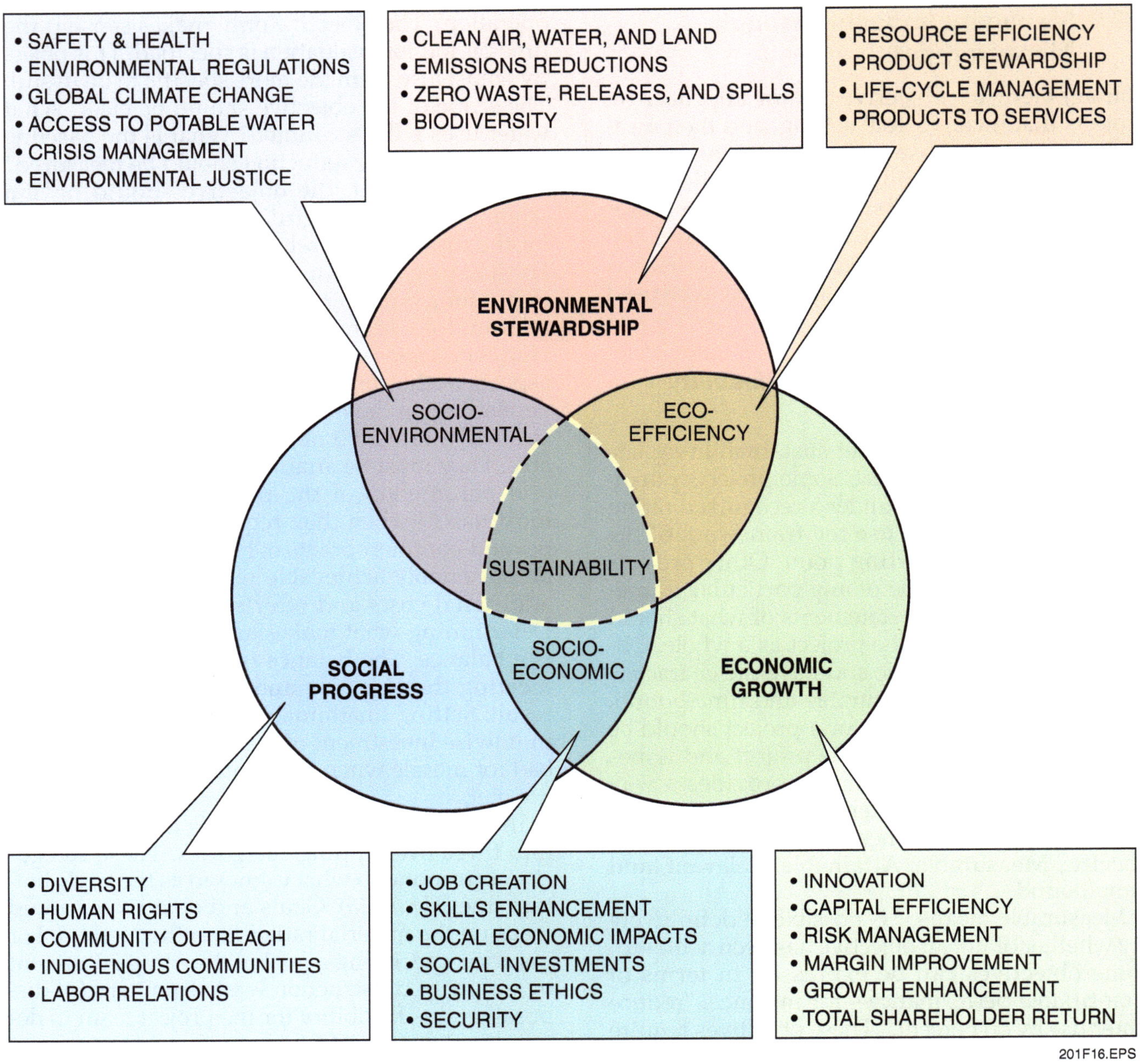

Figure 16 The triple bottom line.

valuable at this stage due to practical knowledge of the job site, as well as an understanding of budget and schedule implications.

After the project team has established sustainability goals and objectives, it is important to agree on how progress will be measured toward those goals. It is also important to determine how the goals will be achieved. Measurement systems for project sustainability provide a structured way to achieve this requirement.

Think About It

Review the factors shown in *Figure 16.* Choose a factor and practice writing goals and objectives associated with that factor that apply during the construction phase of the project.

2.2.2 Measurement Systems for Managing Project Sustainability

Many tools exist to help project decision makers make more sustainable choices. Many of the tools also come with information. These may include resource guides to provide suggestions for improving project sustainability. Some tools result in ratings of projects or materials by a third party. Some provide a means for self-assessment. Still others are used for advertising and marketing purposes. *Figure 17* shows one example of a checklist that can be used to evaluate a construction project.

Of all the tools in the market today, the LEED rating system is probably the most well known for rating and evaluating green buildings. Because of its application to commercial buildings, LEED is the rating system on which this module focuses.

LEED was originally developed by the U.S. Green Building Council (USGBC). The USGBC is a non-profit organization with members from both the public and private sectors. Members include both government agencies and private companies. All aspects of the construction industry are represented. This includes builders, general contractors, architects, engineers, product manufacturers, and others.

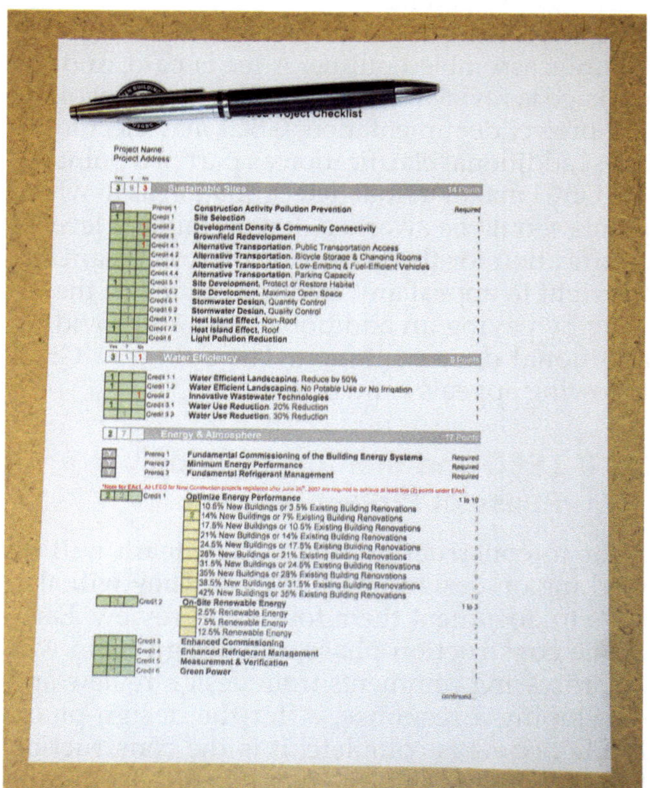

201F17.EPS

Figure 17 Measurement systems, such as LEED, provide criteria and guidance for project sustainability.

In 2008, USGBC created the Green Building Certification Institute (GBCI). GBCI's purpose is to administer LEED certification and professional accreditation. Projects seeking certification under LEED now go through GBCI. GBCI reviews LEED documentation and awards project certification. It also manages the LEED AP and LEED Green Associate exams.

2.2.3 The LEED Green Building Rating System

The LEED rating system applies to a wide variety of project types. These include the following:

- New commercial construction and major renovations
- Existing buildings
- Commercial interiors
- Building cores and shells
- Residential construction
- Neighborhood development

LEED is designed to apply to different climates and contexts throughout the U.S. It has been awarded to buildings internationally as well. LEED has four levels of certification: Certified, Silver, Gold, and Platinum. The LEED rating system consists of a series of performance goals and requirements in six categories, as follows:

- *Sustainable Sites (SS)* – Covers issues related to the location of the project site, impacts to the site during construction, site amenities, and impacts resulting from building operations.
- *Water Efficiency (WE)* – Deals with water consumption and wastewater generation by the building in operation.
- *Energy and Atmosphere (EA)* – Addresses all aspects of the building's energy performance, energy source(s), and atmospheric impacts.
- *Materials and Resources (MR)* – Pertains to the sources and types of materials used on the project, the amount of waste generated, and the degree to which the project makes use of existing buildings.
- *Indoor Environmental Quality (EQ)* – Covers aspects of the building's indoor environment ranging from ventilation to air quality to daylight and views.
- *Innovation in Design (ID)* – Rewards the project for exceeding minimum credit requirements and for using a LEED Accredited Professional (AP).

Each category consists of a series of credits and points that can be earned by a project. Additional points are also rewarded for projects that address regional priorities. Five out of the six categories also have prerequisites. A project must meet all prerequisites in all categories in order to pursue certification (*Figure 18*). It also must obtain a minimum number of **energy performance credits** to meet energy code requirements.

The focus of this module is on the LEED-New Construction Rating System (LEED-NC), Version 3/2009. LEED-NC was the original LEED rating system from which all other versions evolved. It shares general principles, structure, and even some specific credits with all other versions of the rating system. A complete list of the LEED-NC prerequisites, credits, and points is provided in *Appendix A*.

2.2.4 LEED and the Project Delivery Process

The process of certifying a building under LEED has several steps (*Figure 19*). Once the decision is made to proceed with LEED certification of a project, each subsequent step is undertaken using the **LEED Online** documentation system. The first step in the certification process is to register the project with GBCI to declare intent to pursue certification. This allows the project team to access GBCI databases. It also provides an online workspace for the team to manage project documentation. Registering the project requires paying a fee to GBCI that is the same for all projects.

After the project has been registered, the next step is to document green features. These are the features incorporated into the project design that

201F18.EPS

Figure 18 The LEED Green Building Rating System ensures that buildings are constructed to meet or exceed specific environmental standards.

meet credit requirements. Typically, the project team will begin this step by having a meeting. In this meeting they will review the LEED checklist and decide which points and credits to pursue. The process of documentation to prove compliance with LEED credit requirements continues throughout construction. It may also possibly be required for the first year of occupancy, depending on the credits pursued.

The project team can submit documentation to GBCI for review at two points in time: at the end of the design process and after construction is complete. The project team may also opt to submit everything at once when the project is complete. All documentation compiled up to the point of submittal is assembled by the project team using LEED Online. Submittal of documentation for review involves paying a review fee, based on project size. This prompts GBCI to conduct the review. If the team submits a design review, GBCI will evaluate points under the rating system that can be measured at the end of the design process. It will not evaluate credits that require documentation during the construction phase. Design phase review is useful because it gives the project team an idea of how many points they are likely to obtain in the project. It provides a basis for deciding how hard to work for additional points during construction.

At the conclusion of the project, final documentation is assembled online. A fee is paid, and the package is reviewed by GBCI. Upon review of the full project documentation, GBCI may elect to request additional clarification as part of a point audit. GBCI makes a final determination as to which points should be awarded. It determines a level of certification for the project. The project team has the right to appeal any credits declined in the review by paying an additional fee and providing additional documentation. The ruling of GBCI following appeals is final.

2.3.0 LEED and the Construction Phase of Project Delivery

Prior to construction, the design team will be working on design-phase credits. They will also have to document them for design review. Early in the construction phase, the design team will be processing comments from design review and developing a response. After the design-phase LEED review is complete, it is the construction team's responsibility to ensure that these credits are implemented as designed. The construction team is also responsible for construction-phase

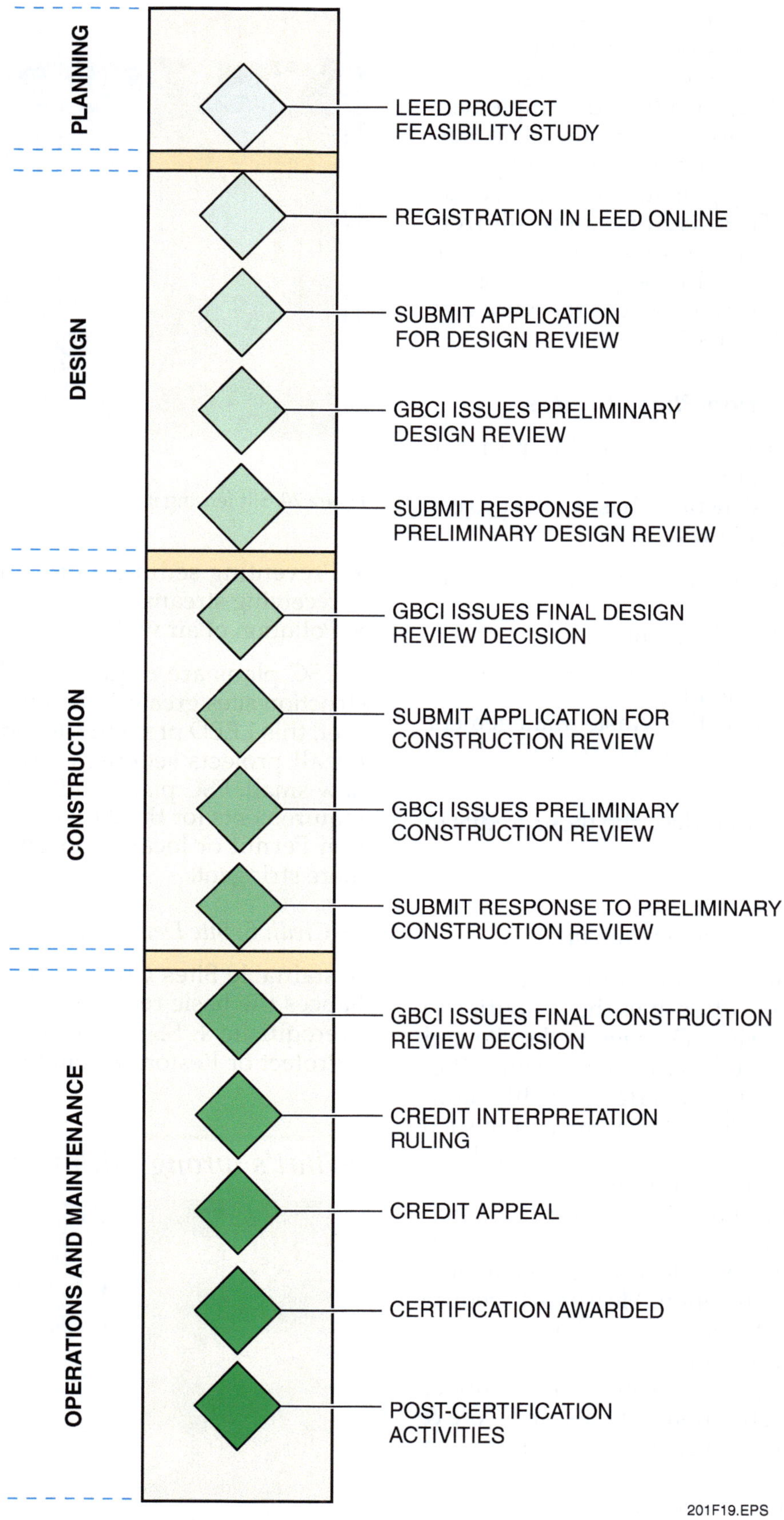

Figure 19 LEED-related activities during project delivery.

credits targeted for the project. A key responsibility of the Sustainable Construction Supervisor includes following through with design-phase credits. It also includes leading the implementation of construction-phase credits. The Sustainable Construction Supervisor also has the opportunity to identify potential actions that can lead to ID credits. These credits can result from substantially exceeding requirements for regular LEED credits. They can also be granted for green activities or project attributes not otherwise covered by regular LEED credits.

2.3.1 Key Construction-Phase Credits

Five construction-phase credits require special attention from the Sustainable Construction Supervisor during the construction phase of the project. These credits are as follows:

- SS Prerequisite 1: Construction Activity Pollution Prevention
- SS Credit 5: Site Development
- EA Prerequisite 1/Credit 3: Fundamental and Enhanced Commissioning
- MR Credit 2: Construction Waste Management
- EQ Credit 3: **Indoor Air Quality (IAQ) Management Plan**

The following sections describe these credits in more detail.

SS Prerequisite 1:
Construction Activity Pollution Prevention

Environmental damage caused to a project site during construction often has the potential to impact neighboring sites. Damage to the site and neighboring sites can take years to overcome. Preventing or minimizing the effects of this damage is critical. Since it focuses on preventing such damage, Sustainable Sites (SS) Prerequisite 1 is a requirement for all LEED projects.

The aim of SS Prerequisite 1 is to reduce pollution from construction activities. This includes controlling soil erosion, waterway sedimentation, and airborne dust generation. Measures to reduce pollution include silt fences (*Figure 20*), temporary plantings, such as grass, and other practices. They are required for all projects being certified under LEED for New Construction, Schools, and Core & Shell. It requires the project to create an **Erosion and Sedimentation Control (ESC) plan** for all construction activities associated with the project. This plan must address the following:

- Loss of soil due to stormwater runoff and wind erosion
- Protecting topsoil for reuse

201F20.EPS

Figure 20 Silt fencing in place to manage stormwater runoff.

- Preventing sedimentation of storm sewers or receiving streams
- Pollution of air with dust or particulates

ESC plans are required by law only for construction sites greater than one acre in size. However, this LEED prerequisite requires an ESC plan for all projects seeking certification, no matter how small. ESC plans under LEED must meet all requirements for the 2003 EPA General Construction Permit or local requirements, whichever is more stringent.

SS Credit 5: Site Development

Sustainable Sites Credit 5 builds upon and enhances the basic requirements established in SS Prerequisite 1. SS Credit 5.1: Site Development – Protect or Restore Habitat is the most relevant

What's wrong with this picture?

201SA04.EPS

for the Sustainable Construction Supervisor. It involves designing and maintaining site operations during construction to minimize damage to the site. SS Credit 5.2: Site Development – Maximize Open Space is more the responsibility of those involved in initial site planning during design. Its focus is on balancing use versus preservation of the site. This balance is between the space taken up by buildings and supporting infrastructure with the amount of open space preserved.

The objective of SS Credit 5.1 is to prevent ecological damage to the project site. If the site has already been damaged by development, the objective is to restore it. It is an option for all projects being certified under LEED for New Construction, Schools, and Core & Shell. One point is available for meeting the credit requirements. An exemplary performance credit may also be obtained if the requirements are greatly exceeded.

This credit is achieved in different ways, depending on whether the site has been previously developed or not. If the site has not been previously developed or disturbed, it is known as a greenfield site. In this case, site disturbance during construction is limited to an area beyond the footprint of the building and other parts of the project, as shown on the **site disturbance plan**. It is the Sustainable Construction Supervisor's responsibility to ensure that construction activities do not extend outside this boundary.

For sites that have been previously developed or graded, the requirements are different. The credit requires restoring or protecting a significant part of the site area to serve as habitat for native species. Vegetated roofs can be included as part of this area. This is because they serve as habitat for birds and insects and can be planted with native plants. Sites that contain both previously disturbed and undeveloped areas require a different approach. A plan should be developed that meets both sets of requirements for the requisite areas of the site.

Meeting the requirements of SS Credit 5.1 requires careful planning and management of construction activities. For example, the area of disturbance should be clearly indicated on the site. This will prevent workers from disturbing protected areas. Contractual provisions and incentives/penalties may be incorporated. Planning of construction operations should include restricted areas for lay-down of materials, recycling and disposal containers, and staging of equipment and materials.

Trees that will be preserved should be protected with fencing at or outside the drip line. This will prevent the compaction of soil within tree root zones by equipment and staged materials (*Figure 21*). The contractor should also carefully plan excavation operations. The goal of planning is to minimize the need for cut and fill during the project.

When restoring previously developed sites, the contractor should also seek opportunities to remove existing paved areas. These areas can be redeveloped as habitat for native species. Opportunities may also exist to recycle pavement materials, either off-site or on-site for another purpose.

EA Prerequisite 1/Credit 3: Fundamental and Enhanced Commissioning

The aim of the commissioning prerequisite and credit under Energy and Atmosphere is to verify that energy-related building systems perform as designed. Commissioning is a systematic process for quality control that evaluates building systems from inception through project delivery. Although outside the scope of LEED-NC, it can also be undertaken after a building is built using the following methods:

- Retro-commissioning (if the building has not previously been commissioned)
- Re-commissioning (if the building has been previously commissioned)
- Continuous commissioning (if commissioning processes are an integral part of owner operations and maintenance)

EA Prerequisite 1: Fundamental Commissioning is required for all projects pursuing certification under LEED for New Construction, Schools,

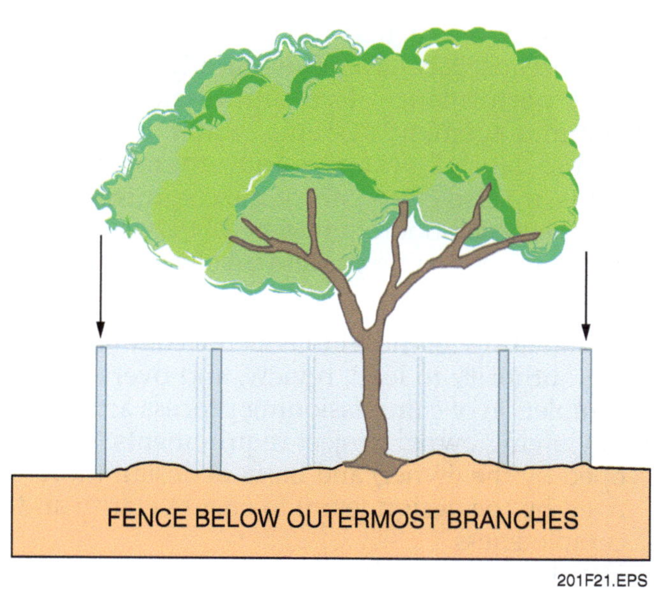

Figure 21 Tree protection fencing helps to preserve existing vegetation.

and Core & Shell. EA Credit 3: **Enhanced Commissioning** is an option for all projects being certified under these LEED systems. One point is available for meeting the requirements of EA Credit 3.

Commissioning should be conducted by a third party not otherwise involved in the project. This will ensure the objective review and evaluation of the building systems. Results must be reported directly to the owner. In some cases, the commissioning agent can be an employee of one of the companies involved in project delivery. This may be the architect, engineering firms, or contractors. In more complex projects, the agent must be an independent consultant or a qualified employee of the owner. Small projects having less than 50,000 gross square feet can use a qualified member of the project team as the commissioning agent. In all cases, the commissioning agent should act independently of the project's design and construction management. LEED requires commissioning agents to have prior commissioning authority experience on at least two other projects in order to be eligible.

At a minimum, the following energy-related systems must be commissioned to meet LEED prerequisite requirements:

- Heating, ventilation, air condition, and refrigeration (HVAC&R) systems, both mechanical and passive, plus associated controls
- Lighting and daylighting controls
- Domestic hot water systems
- On-site renewable energy systems such as photovoltaics or wind turbines

More extensive commissioning efforts may also include systems such as the **building envelope**. This is because it behaves dynamically and affects energy performance. Fire suppression, stormwater management systems, water treatment systems, data/IT systems, and other specialty systems are also often part of the commissioning scope. These are not required by LEED.

The commissioning process involves, at a minimum, the following:

- Designating an individual as the commissioning authority to lead, review, and oversee the completion of commissioning process activities
- Reviewing owner project requirements (developed by the owner) and basis of design (developed by the design team) to ensure clarity and completeness
- Developing commissioning requirements and incorporating them into construction documents
- Developing and implementing a commissioning plan
- Verifying the installation and performance of systems within the scope of the plan
- Completing a summary commissioning report to the owner

To meet LEED requirements for Enhanced Commissioning (EA Credit 3), additional requirements include the following:

- Conducting a commissioning design review prior to the development of construction documents
- Reviewing contractor submittals applicable to systems under the scope of commissioning
- Developing systems manuals for commissioned systems
- Verifying that requirements for training on commissioned systems are met
- Reviewing building operation within 10 months after substantial completion

Many of these requirements should be implemented in conjunction with members of the project team. The commissioning authority is ultimately responsible for reviewing and approving the outcomes developed by the team.

Commissioning requirements are most often described in a specific section of the general conditions of the construction specifications. They may also be referenced on the drawings and in bid forms. They may also be a part of specification sections related to the systems being commissioned.

MR Credit 2: Construction Waste Management

Construction waste is one of the most apparent impacts of the construction process. Recycling construction waste reduces the demand for **virgin materials**. It also reduces the negative impacts of raw material production and limits the damages caused by product disposal. Most construction waste goes to landfills or incineration facilities. The aim of MR Credit 2: Construction Waste Management is to find alternative ways to dispose of such waste. In particular, recyclable materials should be diverted back to the manufacturing process. Reusable materials should be recovered for **salvage** and reuse.

MR Credit 2 is an option for all projects being certified under LEED for New Construction, Schools, and Core & Shell. Up to two points are available for meeting the credit requirements. One point is awarded for diversion of 50 percent of waste by weight or volume. Two points are awarded for diversion of 75 percent of waste. Exemplary performance credit may also be obtained if the requirements are greatly exceeded and 95 percent of waste is diverted.

To achieve this credit, LEED requires a waste management plan. This plan must identify the materials to be diverted from normal disposal. The plan must also describe whether wastes will be sorted on site (*Figure 22*), or comingled. Comingling waste is an option in areas where off-site separation is available. Construction waste management can also be outsourced to companies specializing in diverting the maximum quantity of waste possible.

Excavated soil and land-clearing debris are not counted toward this credit. Achieving the credit requires identifying places to dispose of the various waste streams coming from the project. Contractors working on multiple projects in the same area should share information about possible recyclers for different types of waste.

Ultimately, the environmental benefits of waste recycling must be balanced. The benefits must be balanced against the additional transportation impacts associated with hauling. This is especially true when waste is separated into multiple dumpsters on site. Finding a good location for the dumpsters also requires additional space planning. It may also increase site disturbance. Effective waste management requires careful coordination with the construction schedule. This is to ensure that the minimal number of dumpsters are on site at any given time. It also requires diligent management and education of subcontractors to ensure that dumpsters are not contaminated with wastes that cause them to be rejected by recyclers. Finally, documentation of this credit requires keeping track of waste hauling and disposal activities to provide the necessary information.

201F22.EPS

Figure 22 On-site waste separation for recycling can earn LEED credits.

EQ Credit 3: IAQ Management Plan

Managing indoor air quality (IAQ) during and following construction is an important way to prevent future problems. EQ Credit 3 has two parts. EQ Credit 3.1 focuses on managing IAQ issues during construction. EQ Credit 3.2 focuses on fixing any IAQ problems resulting from construction prior to building occupancy.

Sustainable Construction Supervisors play an active role in achieving EQ Credit 3.1. They must manage the indoor air quality during the construction process. They may also have a role to play in EQ Credit 3.2. This includes scheduling building flush-out or IAQ testing. It also involves ensuring that required environmental conditions are met during these processes.

EQ Credit 3.1 is an option for all projects being certified under LEED for New Construction, Core & Shell, and Schools. EQ Credit 3.2 is only applicable to new construction and schools since it requires interior finishes to be in place. Up to two points are available for meeting both sets of credit requirements.

EQ Credit 3.1 requires the contractor to design and implement an IAQ management plan for the construction and preoccupancy phases of the building. The following key requirements must be part of the plan:

- Meeting or exceeding the recommended control measures developed by the Sheet Metal and Air Conditioning National Contractors Association (SMACNA). This includes installing covers over vents (*Figure 23*). These measures are documented in Chapter 3 of SMACNA's *IAQ Guidelines for Occupied Buildings Under Construction*.
- Protecting absorptive materials and products that are stored on site from moisture damage and other environmental threats.
- Using filtration media with a minimum efficiency reporting value (MERV) of 8 in permanent air handlers used during construction, and replacing those media prior to occupancy.

Smoking is also prohibited inside the building. Additionally, smoking is prohibited within 25 feet of building entrances once the building is enclosed for projects certified under LEED for Schools.

201F23.EPS

Figure 23 Containing construction contaminants is critical for indoor air quality.

To meet the requirements of EQ Credit 3.2, a building flush-out must occur after construction is complete (including punchlist work). Another way to meet the requirements is to complete IAQ testing to prove that the indoor air quality is satisfactory. IAQ testing must be conducted for formaldehyde, particulates, total volatile organic compounds (TVOCs), 4-Phenylcyclohexene (4-PCH), and carbon monoxide (CO). Building flush-outs take time in the construction schedule. These must be at least partially conducted before the building can be occupied. IAQ testing conducted instead of building flush-outs must occur after all interior finishes have been installed. It must also occur prior to occupancy, during normal operating hours and conditions. The Sustainable Construction Supervisor is responsible for scheduling these activities and ensuring that construction does not threaten their completion.

2.3.2 Follow-Through for Design-Phase Credits

In addition to direct responsibility for construction-phase credits, the Sustainable Construction Supervisor is also responsible for verifying that design-phase credits are properly implemented. Materials-related credits from the Materials and Resources category and Credit 4: Low-Emitting Materials in the Indoor Environmental Quality category are particularly important. Other design credits may also require follow-through, depending on project requirements.

MR Credit 3.7:
Materials Reuse and Other Materials-Related Credits

Within the Materials and Resources category, multiple credits can be granted based on using green products and materials (*Figure 24*). These include the following:

- MR Credit 3: Materials Reuse
- MR Credit 4: Recycled Content
- MR Credit 5: Regional Materials
- MR Credit 6: Rapidly Renewable Materials
- MR Credit 7: Certified Wood

The construction team is not directly responsible for material selection during design. However, the Sustainable Construction Supervisor may be asked to contribute to material selection decisions, most often providing information about cost or schedule impacts.

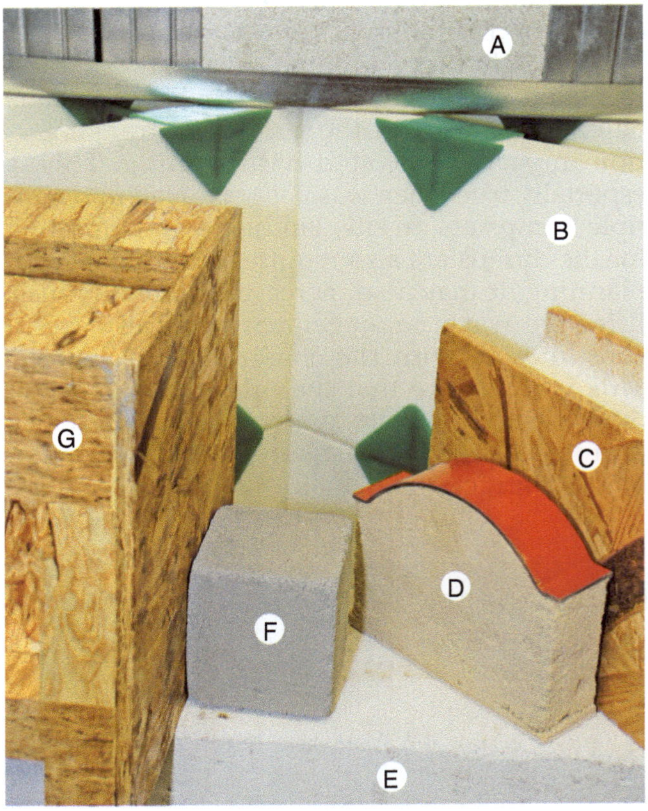

A. Embedded steel stud SIP
B. Polystyrene insulating concrete form
C. Polystyrene stress skin SIP
D. Aerated glass stress skin SIP
E. Aerated autoclaved concrete
F. Fiber-reinforced aerated concrete
G. Straw core stress skin SIP

201F24.EPS

Figure 24 Alternative green structural products.

During construction, the Sustainable Construction Supervisor may contribute to achieving these credits by performing the following tasks:

- Ordering products to meet design specifications
- Collecting documentation to prove that products meet LEED requirements
- Verifying that subcontractors are using compliant products
- Responding to questions from subcontractors about product acceptability
- Making decisions about product substitutions on site
- Verifying compliance during submittal reviews

Each of the listed credits is calculated based on the cost of materials and other product attributes. The Sustainable Construction Supervisor plays a key role in the success of these credits, acting as the project's representative in the field to ensure that any changes do not compromise LEED requirements.

EQ Credit 4: Low-Emitting Materials

Selection and management of materials is also important for EQ Credit 4: Low-Emitting Materials. Sustainable Construction Supervisors should be familiar with relevant certification labels and requirements (*Figure 25*). These requirements will be discussed in more detail in later in this module.

Similar to credits in the Materials and Resources category, the construction team is not directly responsible for material selection during design. However, the Sustainable Construction Supervisor may be asked to contribute to material selection decisions, such as providing information about cost or schedule impacts.

Within EQ Credit 4, Sustainable Construction Supervisors should be aware of product choices and substitutions for all products within a building that have the potential to contribute to indoor air quality through offgassing. These products include the following:

- Adhesives and sealants
- Paints and coatings
- Flooring systems, such as carpet and tile
- Products made from composite wood and agrifiber, such as subfloors, sheathing, and cabinets
- Furniture and furnishings*
- Ceiling and wall systems*

*These products are not included in the LEED rating system. However, they are critical if pursuing certification under LEED for Schools. They contribute to indoor air quality in other types of projects as well. Thus, they are important to consider regardless of whether they are part of formal LEED certification.

Figure 25 Low-emitting paints may be GreenSeal certified.

Other Design-Phase Credits

The Sustainable Construction Supervisor provides quality control for all activities in the field. As a critical member of the project sustainability team, the Sustainable Construction Supervisor must be familiar with the requirements for all design-phase credits selected by the team. Design-phase credits are decided before construction begins. The job of the Sustainable Construction Supervisor is to ensure that what happens in the field matches what was planned. When changes are required in the field, determine whether those changes affect planned design-phase LEED credits. If field changes do affect design-phase LEED credits, bring those changes to the attention of the project sustainability team and document how those changes will influence LEED credit requirements.

2.3.3 Innovation Credits

In addition to design-phase credits and construction-phase credits, there are innovation credits. This is a major area for Sustainable Construction Supervisors to contribute to LEED projects. Innovation credits are used to recognize attributes of projects that are either (a) substantially above and beyond the requirements of ordinary LEED credits; or (b) addressing issues not otherwise covered by existing LEED credits.

Figure 26 shows the Innovation in Design (ID) credit process. The process begins with an idea from someone involved with the project. It may also come from innovation credits used on other projects. Anyone associated with the project can come up with possible ideas. A full list of ID credits is available on the U.S. Green Building Council's website (www.usgbc.org). This is an ideal opportunity for a Sustainable Construction Supervisor to contribute.

ID credits must meet certain basic requirements, including the following:

- Ability to demonstrate quantitative performance improvements with environmental benefits
- Comprehensiveness, which applies to the entire project, and not just a portion of the project or an individual product or system
- Applicability to other projects
- Significantly better than standard sustainable design practices

If the proposed credit does not meet one or more of these requirements, the Green Building Certification Institute will not accept it. This stops the process. If the idea meets all of these requirements, the next step is to develop a draft credit submittal.

The draft of the ID credit submittal must contain certain required components (*Figure 27*). These components illustrate how the project meets the proposed requirements. They also set up the credit in a way that is applicable to other future projects. The LEED Online system provides space for each of these components as part of the submittal.

Many credits involve or relate to the construction process. For example, several innovation credits have been awarded for partnering with

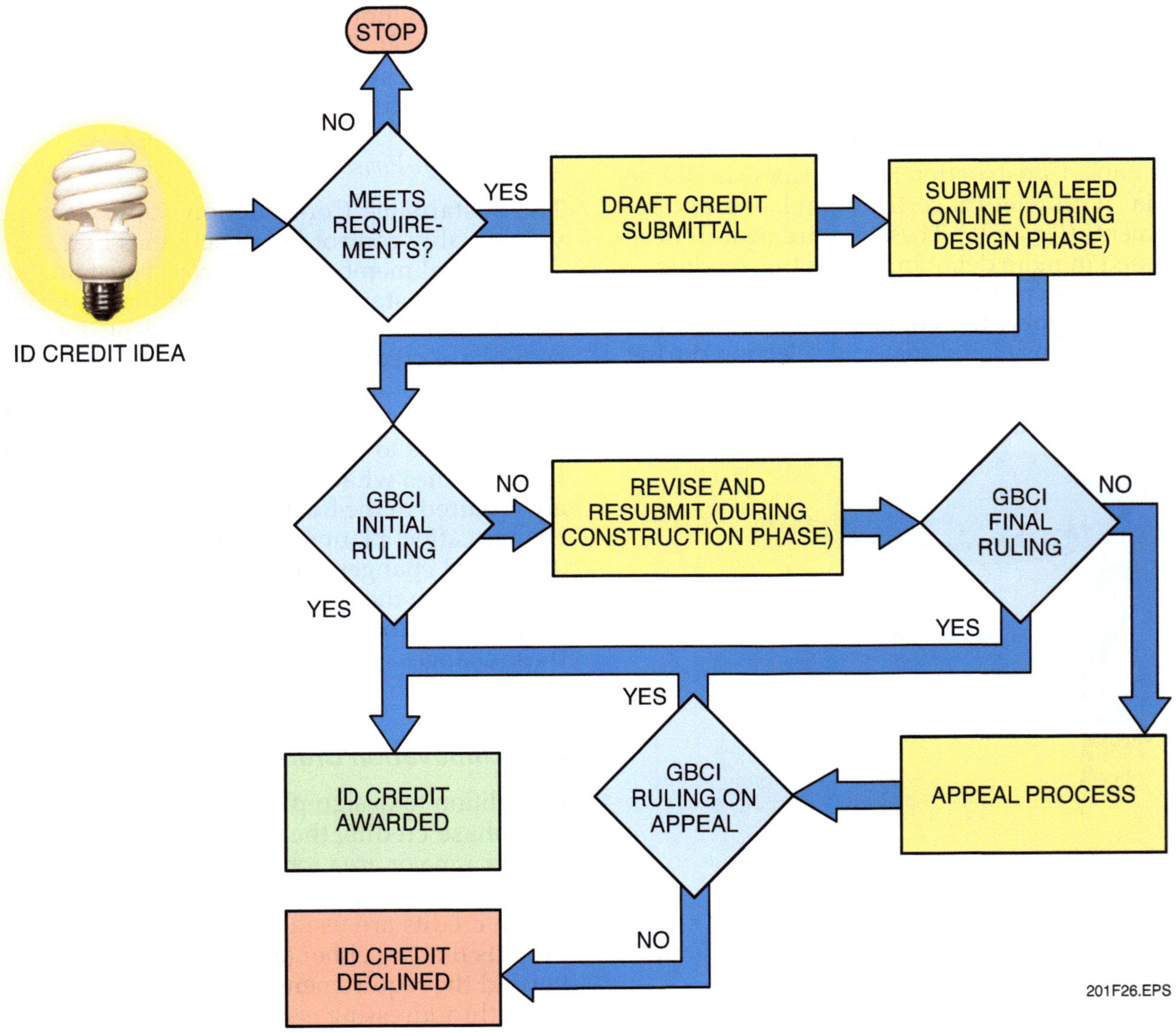

Figure 26 The ID credit process.

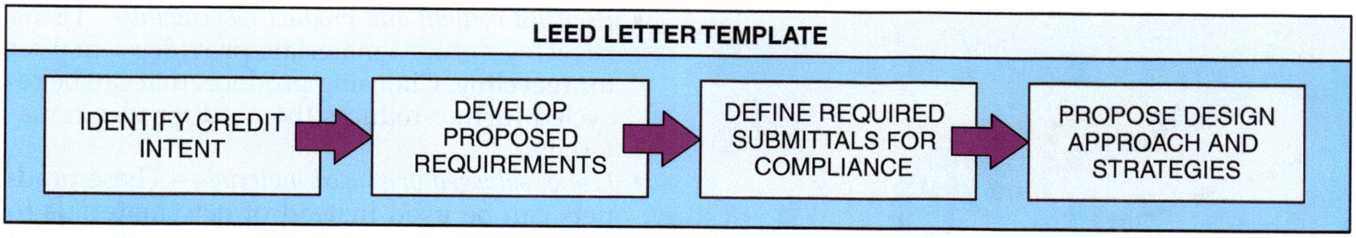

201F27.EPS

Figure 27 Innovation credit elements.

schools, colleges, and universities in various ways. These projects involve students in learning about new building products and systems. They also involve students documenting building performance and learning more about sustainable design (*Figure 28*).

Other construction-related ID credits have been awarded to projects for actions such as the following:

- Reducing site disturbance through use of native raw materials or alternative construction methods
- Remediating the entire project site by removing contamination and restoring habitat
- Rescuing native and adapted plants prior to construction and relocating them on site
- Removing or remediating invasive species on site
- Developing alternative site access and transportation during construction
- Reducing the use of portland cement in concrete mix designs
- Implementing contractor accountability systems for utility monitoring and ongoing building performance
- Providing educational opportunities for the public

201F28.EPS

Figure 28 Involving students can be part of innovation credits.

These and other ideas have been successfully demonstrated on green projects certified under LEED. The project sustainability team should review prior innovation credits. The team should also brainstorm to determine ideas that could be used for these credits.

2.4.0 Managing Sustainability Trade-Offs During Project Implementation

An important consideration for achieving sustainability is making good choices when trade-off opportunities exist. Often the most sustainable choice is not obvious. This is especially true when comparing products or processes that can be used on projects. There are also many opportunities to cause problems during implementation of ideas that were not anticipated during design. The Sustainable Construction Supervisor has an important role to play in reviewing product substitutions and change order requests. Any questions about potential sustainability impacts should be brought to the attention of the project sustainability team for further discussion and resolution.

2.4.1 From Intent to Execution: Potential Pitfalls

Despite the best intentions, things may go wrong during implementation. This is especially true on projects seeking certification under a green building rating system. Common pitfalls include the following:

- *Poor execution of sustainability requirements* – Complex construction projects require the coordination of many parties. Lack of communication, poor understanding of what is required, or lack of motivation to comply can all lead to mistakes in meeting minimum requirements for LEED credits. *Figure 29* is one example of such a problem. Failure to separate construction waste can cause recyclers to reject entire loads of waste. Problems like these must be addressed through quality control and better communication on the job site.

201F29.EPS

Figure 29 Mixed construction waste may not be accepted for recycling in some areas.

- *Poor documentation of sustainability actions* – Projects seeking certification also must meet formal documentation requirements. In a fast-paced project environment, windows of opportunity do not remain open forever. It is critical to keep up with photographic documentation requirements. Systematically organizing receipts and product literature will also pay off in the end.
- *Poor coordination of trades* – Green construction projects often involve new technologies and materials that may be unfamiliar to subcontractors. Even when using conventional materials, tighter tolerances are required to meet energy performance goals for the building. For example, leaving unsealed openings in the floor plate for plumbing or failing to replace wall insulation when disturbed can cause significant loss of building performance.

The Sustainable Construction Supervisor must be on the lookout for these common problems and be sure they are addressed. Otherwise the project's certification could be at risk.

2.4.2 Material Procurement and Product Substitution Trade-Offs

One aspect of the project where trade-offs are especially important is with regard to material procurement and product substitution. Many attributes of materials and products contribute to the overall sustainability of a construction project. These include the following:

- *Recycled content and product recyclability* – Using **recycled content** materials provides a market for recycling. Choosing products that can be recycled further reduces the need for virgin materials.
- *Use of salvaged or reused materials* – These products can be used instead of new materials to conserve resources and also preserve historical character.
- *Use of renewable and rapidly renewable materials* – Using components that can be quickly regrown helps to ensure products will be available on an ongoing basis.
- *Location of manufacture* – The energy required to transport a product is a significant part of its impact. Using local or regional materials reduces this impact while stimulating the local economy (*Figure 30*).
- *Non-toxic, biodegradable materials* – These products are safer to install and promote healthy indoor environments.

Often a product that is more sustainable in one aspect will be less sustainable in another. For instance, sometimes products that are environmentally friendly cost more to purchase. This may

201F30.EPS

Figure 30 Using regional materials is one way to obtain green benefits.

be because they are less widely available. They might be manufactured in only a few locations. Therefore they may have higher transportation requirements. In other cases, non-toxic products may be less harmful to install. However, they might require more extensive maintenance. In both these cases, the good judgment of the project sustainability team is required to determine what is best for the specific case. As the front-line representative of the team in the field, the Sustainable Construction Supervisor must often handle questions about these issues.

2.4.3 Evaluating Requests and Change Order Proposals from Subcontractors

Subcontractors often make requests to do things differently than planned on a project. A subcontractor might have some material left over from a previous project that could be used. Or the subcontractor might want to use a different construction approach than that planned and/or assumed in the project implementation plan. Each of these requests requires careful thought on the part of the Sustainable Construction Supervisor.

Both product and process substitutions can impact the sustainability of a project. In certain critical cases, they can threaten the achievement of LEED credits on the project. For instance, allowing a subcontractor to use leftover products from a previous project may reduce waste, save money, and reduce the need for raw materials. However, if the product contains a higher level of VOCs than permitted by credit requirements, it could jeopardize that credit. This is especially true if indoor air quality testing will be used to demonstrate compliance.

Each request from a subcontractor should be carefully evaluated. Section 3 of this book provides suggestions for evaluating changes in materials during a project. Section 4 provides guidance for evaluating proposed process changes. Please consult these sections for more information on how to evaluate proposed changes of these types.

"I appreciate the opportunity to assist with this meaningful effort. I truly believe that improved field operations, through knowledge and awareness of best practices relative to sustainability, is the next step in making the difference for our environment. The next round of creative and innovative ideas on how to minimize the impact that our projects have on the environment will come from construction professionals in the field that see the work being put in place every day."

Boyd Worsham, LEED AP, Vice President of Construction Support, The Haskell Company

---

Review Questions

1. Construction projects can create negative impacts _____.
   a. inside and outside the site boundary
   b. inside the site boundary
   c. outside the structure
   d. inside the structure

2. Regenerative project sites _____.
   a. create more waste
   b. require more maintenance
   c. consume their own wastes
   d. decrease wildlife habitat

3. For a project to pursue LEED certification it must meet _____.
   a. all pre-requisites in all categories
   b. one pre-requisite for the entire site
   c. the sustainable site pre-requisite only
   d. all the innovation design credits

4. The first step in the LEED certification process is to _____.
   a. declare the intention to pursue certification
   b. consult the USGBC website
   c. document green features of the project
   d. register the project with the USGBC

5. The Erosion and Sedimentation Control plan for construction activities must address _____.
   a. pollution of air with dust
   b. removal of topsoil
   c. construction sequence of operation
   d. construction equipment to be used

6. LEED requires commissioning agents to have prior commissioning authority on at least _____.
   a. one project
   b. two projects
   c. three projects
   d. four projects

7. Developing commissioning requirements and incorporating them into construction documents is _____.
   a. required for LEED certification
   b. required for Energy Star certification
   c. the responsibility of the owner
   d. the responsibility of the contractor

8. Commissioning requirements are most often described in what part of the construction documents?
   a. Site requirements
   b. Soils reports
   c. General conditions of the construction specifications
   d. LEED documentation review

9. To achieve the Materials and Resource Credit 2: Construction Waste Management credit LEED requires a waste management plan that identifies the _____.
   a. materials to be diverted from normal disposal
   b. materials to be landfilled
   c. quantity of land-clearing debris disposal
   d. quantity of excavated soil

10. The Sustainable Construction Supervisor should be aware of product choices for all products in a building that have the potential to impact the IAQ through _____.
    a. aeration
    b. offgassing
    c. particulate emissions
    d. air circulation

11. To ensure proper documentation of sustainability actions, a Sustainable Construction Supervisor should _____.
    a. photograph the site once the project is complete
    b. verbally approve product substitutions
    c. systematically organize receipts and product literature
    d. remove subcontractors from the site who do not comply

12. To prevent a negative impact on LEED credits, requests from subcontractors to change procedures or substitute materials requires _____.
    a. appraisal by the owner
    b. approval by the designer
    c. evaluation of impacts by the contractor
    d. evaluation of impacts by the subcontractor

3.0.0 Green Building Materials and Technologies

One of the most important ways to improve the sustainability of a construction project is by choosing green building materials and technologies. The goal of this section is to introduce the tools needed to ensure that products and technologies used on construction projects are compliant with project sustainability goals and/or LEED requirements. This section describes product labeling and rating systems, sources of product information, documentation for materials and product-related LEED credits, and techniques for evaluating and documenting product substitutions. The main focus of the section is on procurement and documentation tasks. Specifically, this section describes how to do the following:

- Evaluate the attributes of construction products and materials that contribute to their sustainability.
- Find information for products to determine whether they meet a particular standard.
- Know what information needs to be tracked for material- or product-related LEED credits being pursued for a project.
- Document levels of compliance for products and use that information to support basic LEED calculations.
- Manage trade-offs when choosing products to meet project requirements.
- Predict the impact of material substitutions on LEED credits as well as on budget, schedule, and other project execution requirements.
- Effectively bring the implications of material selection to the attention of project decision makers.

3.1.0 The Role of Materials and Technologies in Project Sustainability

Materials and building systems strongly influence the sustainability of the resulting project. These products have an impact as part of the building. They also have upstream and **downstream impacts** resulting from their manufacture, transport, and disposal. As components within a larger building, materials and technologies influence the performance of the building and the quality of the indoor environment. They also impact the health and safety of construction workers who install them. Finally, materials have a broad range of impacts called **externalities**. These are impacts that are not directly felt by the building or its stakeholders. Instead, these impacts are borne by society as a whole. They are affected by a material's recycled content, **regionality**, rapid renewability, and **certified sustainable harvest**.

3.1.1 Materials and Climate

Construction materials have a direct effect on the local environment and climate where they are installed. They have the ability to absorb or reflect heat, which affects building heating and cooling loads as well as temperatures near the building itself.

The **albedo** of a material is the amount of energy that is reflected back when light or solar radiation hits the surface. In conjunction with thermal mass, albedo determines how comfortable it is to stand on a piece of pavement on a sunny day (*Figure 31*). High-albedo pavements tend to reflect solar energy and do not absorb heat as readily as dark pavements. Dark pavement contributes to the **urban heat island effect**.

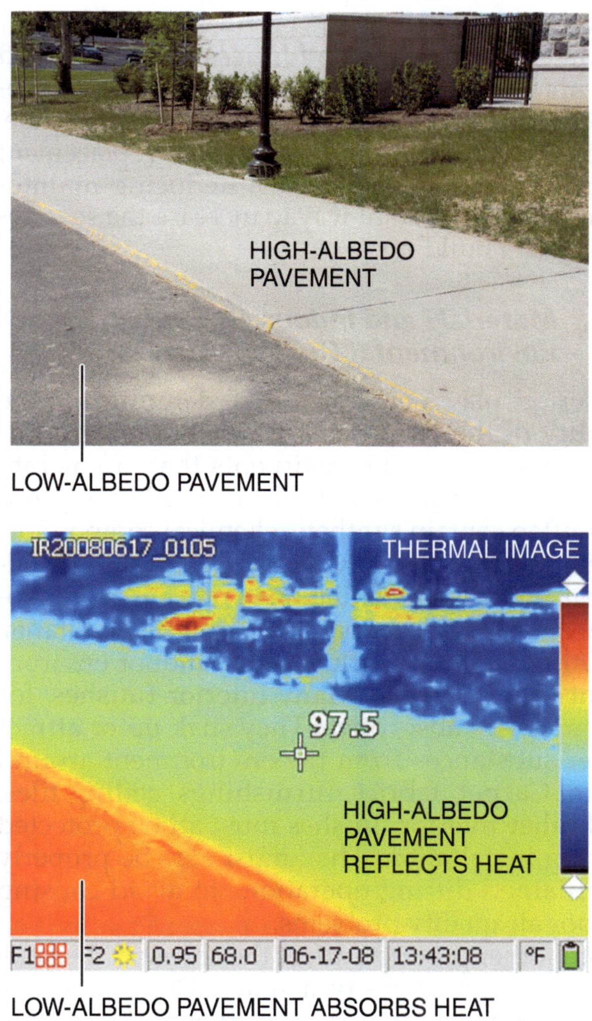

Figure 31 Urban heat islands are worsened by dark pavements.

The same is true of materials used for other parts of the building envelope. Reflective coatings are available for roofs that reduce the amount of heat gained by the building. These coatings help to reduce the cost of air conditioning the spaces inside the building during hot days.

3.1.2 Materials and Building Performance

Materials affect building performance in other ways as well. All materials are chosen during design to play a specific function within the building. These roles may be aesthetic, structural, or some other role. Some materials play an especially critical role in the building's energy performance (*Figure 32*). Products used in the building envelope, including exterior walls, roof, windows, and doors, must work together to provide a barrier between the outside and indoor environments. Improper installation of materials, especially these critical components, can have a lasting effect on how well the building works and how much it costs to operate.

Material durability and fitness for its intended use are also important. Green products are sometimes selected because they will be more durable. They may also require less frequent replacement than conventional materials. Reducing maintenance is an important way to increase the sustainability of a building.

3.1.3 Materials and Indoor Environmental Quality

Materials play a critical role in determining the quality of the indoor environment. Some products have chemical constituents that offgas into the indoor environment (*Figure 33*). These products often contain synthetic chemical compounds. They must be carefully managed through installation and proper ventilation to prevent overexposure of building occupants to airborne pollutants.

Other materials influence the indoor environment in other ways. Many interior finishes, for example, are absorptive. They soak up or attract other substances from the environment around them. Carpet, fabrics, furnishings, ceiling tiles, and other interior finishes must all be protected during installation. They must also be properly maintained during operations to avoid causing indoor air quality problems.

Still other products affect the acoustic, thermal, or visual properties of a space. Designers take great care to choose materials that will influence how people experience a building. All of these properties contribute to a material's influence on environmental quality.

201F32.EPS

Figure 32 Improperly installed insulation can affect building performance. This insulation does not extend the full height of the wall cavity.

201F33.EPS

Figure 33 Interior finishes may emit VOCs and reduce IAQ.

3.1.4 Materials and Construction Safety

Construction materials also have the ability to affect the health and safety of workers who install them. For instance, masonry, concrete, and some aggregate products all contain silica particles that can become airborne during handling (*Figure 34*). Workers who do not take appropriate precautions or use proper methods to avoid these problems can suffer long-term health impacts. The same can be said for many other products as well.

201F34.EPS

Figure 34 Dry-cutting cementitious products can lead to silicosis. Appropriate respiratory protection must be worn.

3.1.5 Materials and Externalities

Building materials have direct impacts on the occupants of a building and the workers who construct it. Building materials also have an influence on society at large. These are called externalities because their effects are felt external to the project itself.

One important externality is the influence a material has on the total stock of resources available to meet human needs. For instance, many building products are made or transported using non-renewable resources such as fossil fuels. Using these products causes the total stock of fossil fuels available for use to be reduced. This reduces the ability of others to use those fuels for other purposes. While the amount of fossil fuels used in the manufacture of one material for a single building might be quite small, the cumulative effect from all the materials used for construction is quite large.

Using a product with recycled content is one way to reduce the externalities of that product. As described earlier, recycled content material generally requires less energy to produce than virgin content material. It has lower **embodied energy**. This means that more energy is available for other purposes or for future generations.

The recycled content in a building product can come from two different sources: pre-consumer and post-consumer. **Pre-consumer recycled content** comes from waste generated during the manufacture of the material. Using this waste to make new materials is better than disposing of it. However, ultimately it represents inefficiency in the manufacturing process that should be engineered out of that process.

Post-consumer recycled content comes from products after they have been used by the consumer and returned for recycling. The amount of recycled content is often stated on the product's label. Recycled content can be certified by a third-party organization so that consumers know the manufacturer's claim is authentic (*Figure 35*).

Another attribute of materials that affects externalities is regionality, which is the degree to which a product has been manufactured near the point of use. Using regional materials reduces the energy required for transport, minimizes the pollution resulting from that transport, and increases the health of the local economy by supporting local or regional businesses.

A third attribute of building materials that influences society at large is rapid renewability. A material is rapidly renewable if its components can be regrown quickly without damage to the environment. Many building materials are becoming available that are made from rapidly renewable components. *Figure 36* shows some of the common agricultural products that are used as part of rapidly renewable building materials.

Finally, some building materials are certified as sustainably harvested. This means that they were harvested from the environment in a way that could be continued indefinitely without harm. *Figure 37* shows a **Forest Stewardship Council (FSC) Certified** wood label. These materials may be labeled by a third-party certification organization that reviews the harvest practices to ensure they are sustainable.

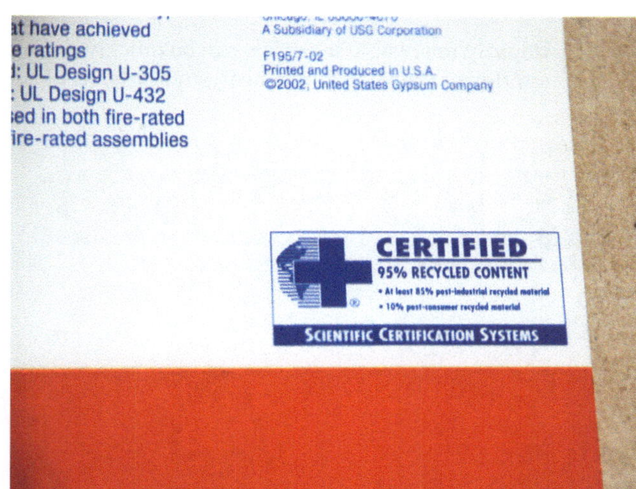

201F35.EPS

Figure 35 Certified recycled content label. This product has been certified by Scientific Certification Systems to have 95 percent recycled content.

Figure 36 Rapidly renewable materials can be quickly regrown from various sources such as (from top, clockwise) sunflower seeds, cotton, wheat, soybeans, wool, and bamboo.

Figure 37 FSC Certified wood label.

These are just a few of the attributes of building products that reflect how they impact society. Others include how biodegradable a product is, how toxic its components are, and how its manufacture affects the health of natural ecosystems.

3.2.0 Green Product Rating and Labeling Systems

Green product rating and labeling systems are one way to make sense of the many material attributes affecting product sustainability. Many sources have developed information that helps building stakeholders understand how materials will perform. This information also identifies what impacts they have.

3.2.1 Types of Green Product Identification/ Evaluation Systems

Green product information is available in a variety of forms. For some attributes, such as recycled content or sustainable harvest, certification systems are available. These systems allow third parties to review and objectively confirm a manufacturer's claim. Certification systems usually have multiple criteria. These criteria must be met in order for a product to be certified.

Multiple directories of green products are also available. Some directories may allow product manufacturers to suggest products to be listed. These directories function like yellow pages. They do not necessarily have any third-party review or qualification of what products are listed. Other directories, such as the online materials database at www.buildinggreen.com, provide screening of products before they are listed. A product's qualifications are explicitly listed in the BuildingGreen database.

Still other directories are available in print form (*Figure 38*). These guides can quickly become outdated as new products enter the market. However, they are still a good starting point. R.S. Means offers cost information on green products in its guide, *Green Building: Project Planning & Cost Estimating*. Both *GreenSpec* and *Green Building Materials* by Spiegel and Meadows contain examples of specification language that can be used to ensure sustainable products are used in a project. *GreenSpec* also includes listings of products by manufacturer. It is the print version of the BuildingGreen.com database. The *Handbook of Sustainable Building, Green Building Handbook,* and *Environmental Resource Guide* all contain more general information about the environmental performance and life cycle impacts of building materials. They do not provide specific manufacturer or brand information.

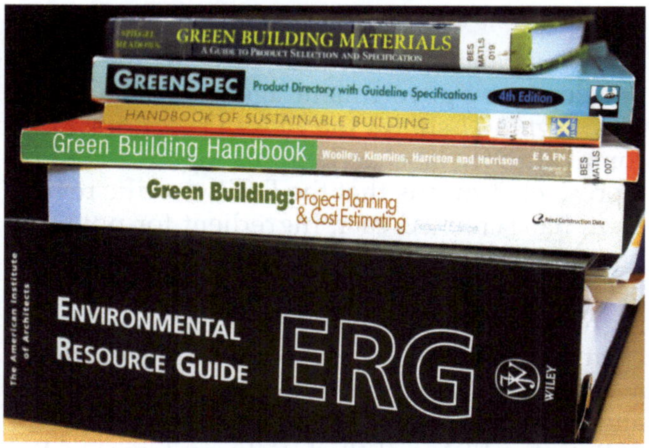

201F38.EPS

Figure 38 A variety of directories of green products are available.

Product labeling systems also provide various kinds of information about a product's green attributes. Labels may provide values for specific qualities about a product (for example, grams per liter of volatile organic compounds or percent recycled content). Or, they may display a logo that indicates the product has met some requirement.

Life cycle assessment is a still more detailed set of information about a product. It is used as the basis for some green product certifications. Resources, such as the *Environmental Resource Guide,* contain life cycle assessment information about building products from their manufacture to end-of-life-cycle (*Figure 39*). This can also be called cradle-to-grave analysis.

3.2.2 Certifying and Standards Development Organizations

Several organizations provide certification of standards or rating systems. These include the following:

- *American National Standards Institute (ANSI)* – This organization (www.ansi.org) creates guidelines and standards that apply to many industries including construction. ANSI standards are often applied to building assemblies. ANSI also accredits standards developed by other organizations. They do this to ensure that they conform to international requirements. For instance, the LEED Green Building Rating System is accredited by ANSI.
- *American Society of Testing and Materials (ASTM) International* – This organization (www.astm.org) develops standard test procedures for various types of materials and processes. For instance, ASTM International has developed *ASTM C-920*, a standard specification for the properties of elastomeric joint sealants (*Figure 40*).
- *International Organization for Standardization (ISO)* – This organization (www.iso.org) develops and maintains international standards. It is a network of national standards organizations from 161 countries. *ISO 14000: Environmental Management Systems* is an example of a standard used by some companies in the construction industry to improve their environmental performance.

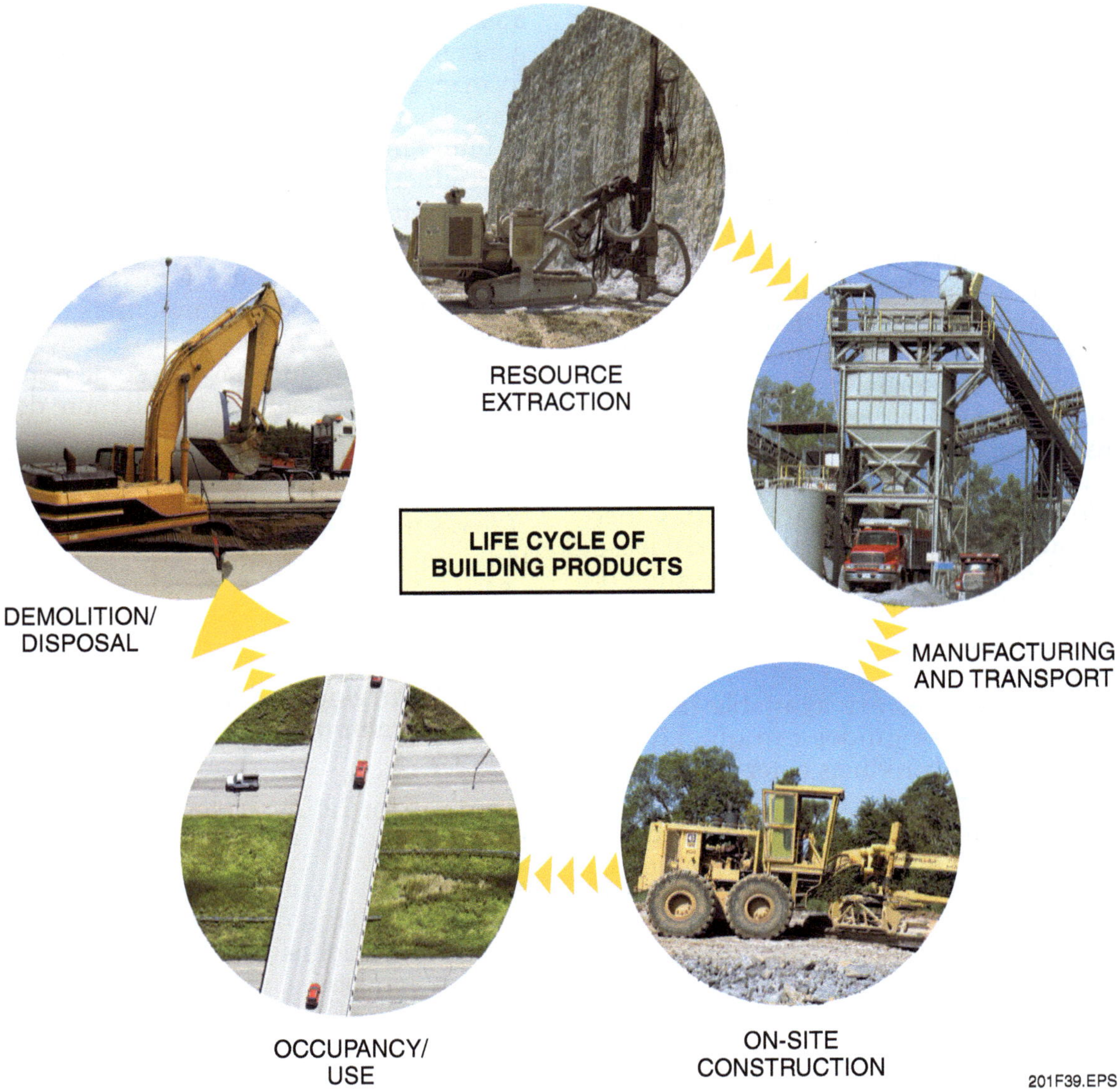

201F39.EPS

Figure 39 Life cycle assessment takes into account all impacts necessary to create, use, and dispose of a product.

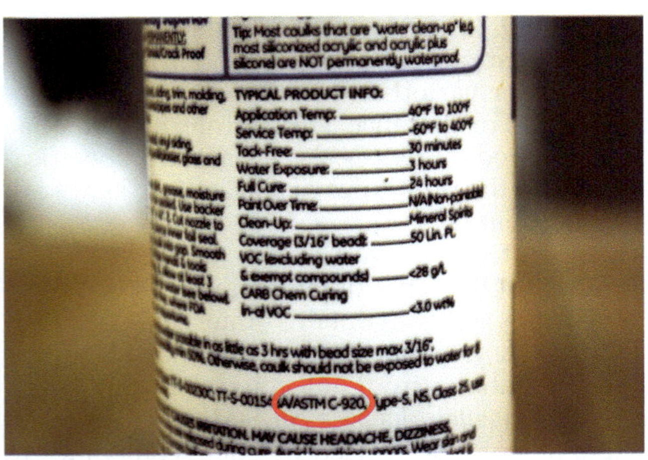

201F40.EPS

Figure 40 Standards development and management organizations provide testing standards.

Other standards or rating systems have been developed by government agencies such as the EPA. *Figure 41* shows the EPA's Design for Environment (DfE) label. When this logo is found on a product, it means that the DfE scientific review team has screened each ingredient for potential human health and environmental effects. The label indicates that the product contains only those ingredients that pose the least concern among chemicals in their class. It is based upon currently available information, EPA predictive models, and expert judgment.

Both the EPA and the U.S. Department of Energy (DOE) were original sponsors of the LEED Green Building Rating System. This system has since become the custody of the U.S. Green Building Council and the Green Building Certification Institute.

201F41.EPS

Figure 41 The EPA develops product and building standards. This product meets EPA's Design for the Environment criteria.

Some standards are also developed by private sector organizations. These organizations are most commonly non-profit organizations. Some examples are:

- *Scientific Certification Systems (SCS)* – This organization (www.scscertified.com) has the capability to certify a variety of manufacturer environmental claims (*Figure 42*). It is most well known for certifying products as being sustainably harvested under the Forest Stewardship Council's rating system.
- *Green Seal* – This organization (www.greenseal.org) has developed standards for over 40 different types of products ranging from paint to coffee filters. It is an independent, third-party certification and standards development agency. This agency evaluates the environmental properties of many types of products, including building materials such as paint, and housekeeping chemicals.
- *McDonough Braungart Design Chemistry (MBDC)* – This organization (www.mbdc.com) provides sustainability consulting and product certification under the **Cradle2Cradle** label. Products certified under Cradle2Cradle have been evaluated for impacts on human health, environmental health, and **recyclability**.

3.2.3 Common Green Rating/Labeling Standards

A variety of green product labels can be found on materials used in green building. Some of the more common labels are described here. There are many others that are not as well recognized.

201F42.EPS

Figure 42 SCS is an independent third-party evaluator. For this product, SCS has certified the level of recycled wood content.

The GreenSeal logo (*Figure 43*) is awarded to products that meet requirements developed by the GreenSeal organization. It is based upon environmental or human health performance. The GreenSeal logo is awarded to products for meeting indoor air quality standards. *GreenSeal Standard GS-11: Paints and Coatings* is a reference standard under the LEED rating system. For instance, all

201F43.EPS

Figure 43 The GreenSeal logo for products meeting IAQ standards.

interior paints and coatings used in a project must be GreenSeal certified in order to receive a point under EQ Credit 4.2. GreenSeal also certifies the environmental and energy performance of windows, occupancy sensors, cleaning chemicals and services, and other building-related products.

Some product labels indicate a product's recycled content (*Figure 44*). Many products make claims of recycled content. However, these claims should be third-party certified by a reputable organization, such as SCS. Product labels should also indicate the source for the recycled content (pre- or post-consumer/recovered). Most often, recycled content is expressed in terms of unit weight of the product.

Products can also be labeled to indicate their recyclability. The recycle symbol (*Figure 45*) is often included on product labels when some portion of the product or its packaging is recyclable. In general, this symbol can be used on any product or package that contains content that is technically recyclable. However, the presence of the logo does not mean that the whole product or package is recyclable. It also does not indicate that recycling facilities are available to take the product. For products displaying the recycle symbol, be sure to consider what parts of the product and/or its packaging are actually recyclable. Also consider

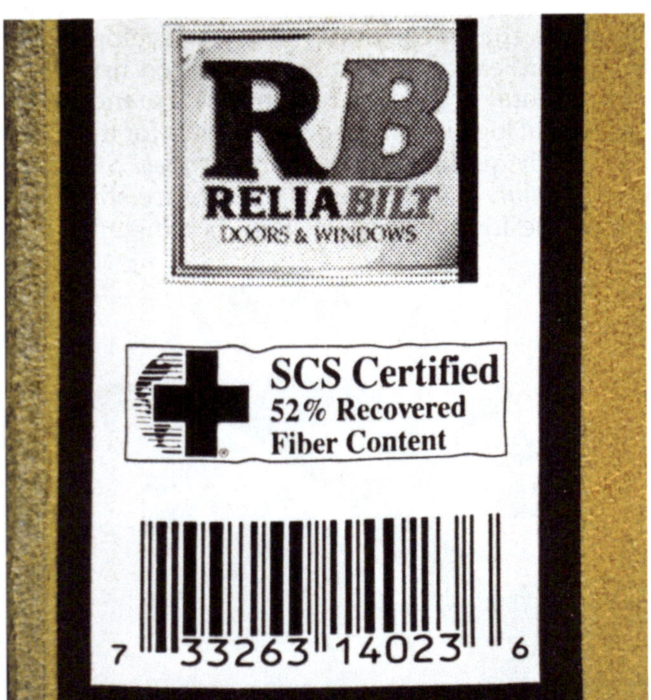

201F44.EPS

Figure 44 The SCS recycled content logo. This door contains 52 percent recovered fiber.

201F45.EPS

Figure 45 This label indicates a variety of environmental features of the product, including the standard symbol indicating that the product is recyclable at the bottom of the label.

whether recycling facilities are available in the area to take that type of material.

The **Energy Star** logo (www.energystar.gov) is another common label used to indicate that a product has an environmental performance benefit. The EPA originally developed the Energy Star label. It is applicable to appliances, building products, home electronics, office equipment, and whole buildings. The EPA provides technical reviews of product attributes before awarding permission for a product to display the Energy Star logo (*Figure 46*). Products must perform in the top 25 percent of all products within their class in order to display the Energy Star logo.

WaterSense (www.epa.gov/watersense) is a program developed by the EPA to indicate water-conserving devices. WaterSense applies to residential and commercial plumbing products. All products displaying the WaterSense logo (*Figure 47*) have been third-party tested and certified to

201F46.EPS

Figure 46 The EPA Energy Star logo for energy-efficient products and buildings.

meet EPA standards. WaterSense specifications are available for toilets, urinals, water faucets, showerheads, and new homes. New standards are under development for irrigation fixtures and other water-consuming devices. Rebates are also available for WaterSense products.

201F47.EPS

Figure 47 The EPA WaterSense logo for water-efficient products.

GREENGUARD (www.greenguard.org) is a label (*Figure 48*) indicating that a product meets strict chemical **emissions** standards and therefore contributes to healthier indoor air quality. GREENGUARD is awarded by the GREENGUARD Environmental Institute. This is an independent third-party certification organization. Products certified under GREENGUARD include building products, such as adhesives, cleaning and maintenance products, and furniture. GREENGUARD provides a reference standard under the LEED rating system to establish the indoor air quality impacts of furnishings.

The **CRI Green Label** and Green Label Plus (www.carpet-rug.org) are standards that apply to carpets, carpet adhesives, and carpet pads.

201F48.EPS

Figure 48 The GREENGUARD certification logo for products meeting IAQ criteria.

Developed by the Carpet and Rug Institute, a trade association for the carpet industry, this logo (*Figure 49*) indicates that a product has been tested to meet emissions standards for multiple chemicals including benzene, formaldehyde, and others. The GREENGUARD standard is a reference standard for flooring products under the LEED rating system.

Logos are also used to indicate that wood products have been certified sustainably harvested under standards established by the Forest Stewardship Council (FSC). See *Figure 50.* SCS is the organization that provides third-party certification of claims under FSC standards. Products meeting FSC (www.fsc.org) requirements must have chain-of-custody certification from initial harvest of wood products through final manufacture.

The **National Sanitation Foundation (NSF)** develops national standards. It also provides third-party conformity assessment services for building products and other types of products and services. NSF (www.nsf.org) is best known for certifying plumbing system components, including piping, valves and fittings, and treatment system components, such as filters (*Figure 51*). NSF also provides assessment of products using methods such as eco-efficiency analysis.

3.3.0 Sources of Product Information

The product labeling systems described here represent a variety of **eco-labels** on the market today. They are, however, by no means comprehensive. Rating systems are constantly being updated. New logos are regularly appearing on the market. Some labels are audited or certified by third-party organizations. Others may simply represent unsubstantiated manufacturer claims. Additional research may be necessary to determine which are credible and which are not.

201F49.EPS

Figure 49 The Carpet and Rug Institute's Green Label Plus logo for certified carpet.

201F50.EPS

Figure 50 The FSC label for certified wood. SCS provides third-party verification for this label.

201F51.EPS

Figure 51 The NSF product certification label.

3.3.1 Evaluating Manufacturer Claims

Figure 52 shows an environment-related claim on a product label. It shows the types of information provided to consumers about a product's environmental performance. In this case, the label does not make a product-specific claim. It quotes a government agency about how general energy-saving practices can influence heating and cooling costs. By reference, the implication is that using this product can provide similar performance. There is no specific proof or evidence provided.

Messages such as this one can be confusing to the uninformed consumer. Since the label does not make any product-specific claims about energy savings, it is technically correct. However, it does not provide the necessary details to evaluate what the product itself can do.

Still other product labels and logos appear on product packaging without any reference to an established standard. The EarthSmart™ label has been trademarked as a graphic and is used on the labels of various building products (*Figure 53*). However, it is not linked with a recognized standard and does not require any testing or verification prior to use. As such, it is not a reliable basis for decision-making during product selection.

These examples show how labels that may initially suggest good environmental performance can fail to provide the information needed for decision-making. Although both of the products bearing these labels may have environmental benefits, the label does not provide enough objective, verifiable information to effectively evaluate the product.

201F52.EPS

Figure 52 Manufacturer labels can be confusing. This label makes an environmental marketing claim backed up by a general web reference.

201F53.EPS

Figure 53 Not all logos indicate compliance with a standard. This label is not based on a recognized standard.

3.3.2 Reading Product Labels and Other Documentation

Product labels can contain a wealth of information. They can also contain claims without a sound basis in scientific measurement. Misleading claims about a product's environmental benefits are known as **greenwash**. Greenwash has been defined as the act of misleading consumers regarding the environmental practices of a company or the environmental benefits of a product or service. TerraChoice, a firm specializing in environmental marketing, has identified several common types of greenwashing:

- *Hidden Trade-Offs* – A claim based on a small number of environmental attributes without attention to other important environmental issues.
- *No Proof* – A claim that cannot be supported with easily accessible data or reliable third-party certification.
- *Vagueness* – A claim that is so broad that its meaning is likely to be misunderstood by the consumer. All-natural is a good example—mercury and formaldehyde are naturally occurring but not necessarily green.

- *Irrelevance* – A claim that may be truthful but is unimportant for decision-making. CFC-free is a good example—CFCs are banned by law, so nearly all products are CFC-free.
- *Lesser of Two Evils* – A claim that may be true within the product category but distracts the consumer from other negative product attributes. Organic cigarettes are a good example.
- *Fibbing* – A claim that is just plain false. While infrequent, this is still a problem in some cases.
- *False Labels* – A claim made using a made-up or fake eco-logo or claims of third-party endorsement where none exists.

Be on the lookout for these problems when evaluating manufacturer claims. Even if product label information is completely accurate, it still may be difficult to find relevant information on a product label. *Figure 54* is one example of a label design that makes finding relevant information difficult.

The best way to approach reading a product label or manufacturing information is to know what types of attributes are most relevant for that type of product. The LEED rating system provides a good starting point.

3.3.3 Directories and Databases

Another good source of objective product information is through online or print databases. *Figure 55* shows examples of some of the different types of material comparisons found in leading print references on green building materials.

When reviewing information about a product, consider the following:

- What is the source of the information? What is its interest in the product? Information from independent sources (such as universities or government organizations) may be more reliable than information from sources with a vested interest in the product (such as trade associations).
- What is the basis of the information? Is it supported by third-party verification, or is it provided by the product manufacturer? Is contact or reference information provided for the third-party verifier?
- Are standardized tests from recognized agencies, such as ASTM or ANSI, used to produce data about the product? Using standardized test methods to evaluate products allows them to be compared with other similar products.

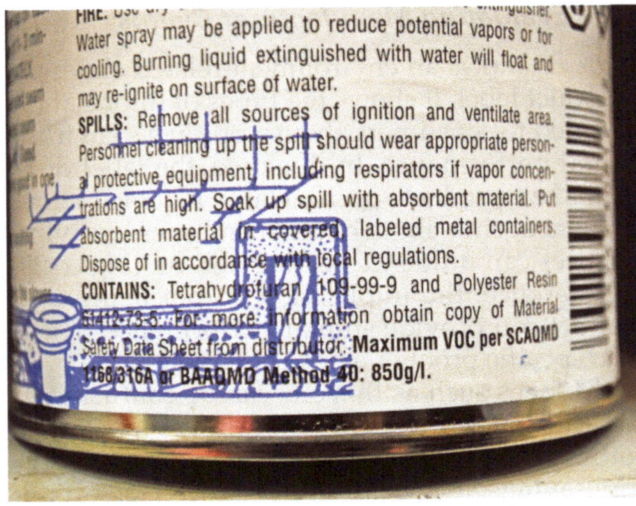

201F54.EPS

Figure 54 VOC content for products is generally specified on the label, expressed as grams per liter (g/l). In this case, the information is mostly obscured by the product label design.

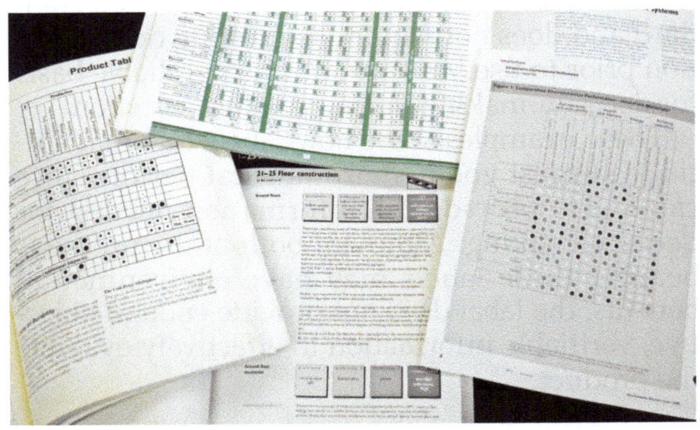

201F55.EPS

Figure 55 Product comparisons are available in various directories.

- If the product has an eco-label, is the label from a recognized, credible, third-party source? Labels such as the one shown in *Figure 53* may appear to be supportive of environmental benefits, but there are no limitations on their use and they are not based on an objective standard of review.
- Is extraneous information provided that is vague, irrelevant, or distracting? The label shown in *Figure 52* is an example of this type of information. It may lead to conclusions about the product that are not substantiated in fact.

3.3.4 *Asking Effective Questions of Manufacturers, Vendors, and Suppliers*

Obtaining information about a product's environmental benefits or sustainability can be challenging. Valuable sources of this information include product manufacturers, vendors, and suppliers.

The checklist of Questions for Product Manufacturers, Vendors, and Suppliers provides a set of questions necessary to obtain information for LEED credits. When collecting product information, Sustainable Construction Supervisors should coordinate efforts with their company and other members of the project sustainability team. For commonly used products, this information can be collected and archived for use in future projects and LEED calculations. This can save considerable time on subsequent projects.

3.4.0 Calculating and Documenting Materials-Related LEED Credits

The Sustainable Construction Supervisor may play various roles in the LEED certification process. Both design- and construction-phase credits depend on proper documentation. For materials-related credits, documentation obtained during construction is required to validate assumptions made during design. Materials-related credits requiring attention from the Sustainable Construction Supervisor include the following:

- MR Credit 3: Materials Reuse
- MR Credit 4: Recycled Content Materials
- MR Credit 5: Regional Materials
- MR Credit 6: Rapidly Renewable Materials
- MR Credit 7: Certified Wood
- EQ Credit 4: Low-Emitting Materials

Many of these credits require that similar types of information be collected about the product, as shown in the checklist for Materials-Related LEED Credits. The specific responsibilities of the Sustainable Construction Supervisor will vary from project to project and company to company. Be prepared to function in any of the listed roles.

The following sections describe LEED certification and the documentation required for these credits. All of the MR credits described here are calculated based on the material cost of qualifying products versus total material cost for the

CHECKLIST
Questions for Product Manufacturers, Vendors, and Suppliers

- ❑ Where is the product made?
- ❑ Where are the raw materials from?
- ❑ What environmental claims does the manufacturer make?
  - Recycled content
  - Recyclability
  - Rapid renewability
  - Certified sustainable harvest
  - Non-toxic
  - Biodegradable
  - Low-emission
- ❑ Are these claims backed up by third-party certification?
- ❑ What organization certifies the product or verifies manufacturer claims?
- ❑ What documentation is available? Where can it be obtained?
- ❑ Are there any special certifications or training required to install the product and maintain the warranty?
- ❑ Any there any special requirements for purchase, shipping, storage, or installation?
- ❑ Are there any comparable alternative products? How do they compare?

CHECKLIST
Materials-Related LEED Credits

- ❑ Review the project specs to identify special product requirements
- ❑ Procure products to meet specifications
- ❑ Obtain documentation from vendors, suppliers, or manufacturers to verify and support LEED compliance
  - Actual material cost (product, delivery, taxes and fees)
  - Replacement cost (MR Credit 3)
  - Pre- and post-consumer recycled content (MR Credit 4)
  - Location of manufacture and raw materials (MR Credit 5)
  - Rapid renewability (MR Credit 6)
  - Certified sustainable harvest (MR Credit 7)
  - Low-emitting certifications (EQ Credit 4)
- ❑ Complete or verify LEED calculations for products
- ❑ Respond to subcontractor requests for product substitution
- ❑ Review change order requests
- ❑ Bring potential problems or changes to the attention of the project sustainability team
- ❑ Verify correct product use by subcontractors and installers

project. The exception is MR Credit 3: Materials Reuse. The calculation can be based on either the replacement cost or the actual cost of salvaged materials used in the project. A key responsibility of the Sustainable Construction Supervisor is to document material costs for use in the calculations.

Calculations for MR credits also require an estimate of the total material costs for the project. This can be precisely calculated if material costs are known. It can also be estimated as 45 percent of the total construction cost of the project. Depending on the nature of the project, it can be beneficial to use actual material costs versus estimated material costs. Material costs for the project include all expenses to deliver the material to the project site. They include taxes and transportation costs incurred by the contractor. They do not, however, include labor and equipment once the material has been delivered to the site.

EQ Credit 4: Low-Emitting Materials, on the other hand, is based on having all applicable products under each category be compliant with requirements. EQ Credit 4 can also be earned by demonstrating that the project uses a combination of products that collectively keep it under the required VOC levels. For these credits, the cost of the materials is not required for the calculation. However, it will still be required as part of the total material cost calculation for any MR credits being pursued. In the case of EQ Credit 4, the main role of the Sustainable Construction Supervisor is to verify that compliant products are being used.

3.4.1 MR Credit 3: Materials Reuse

MR Credit 3 encourages the use of salvaged or reused materials as part of a construction project. Use of these materials has environmental benefits. It reduces the need to dispose of products in a landfill. It also reduces the energy and materials needed to manufacture new products. Up to two points can be obtained under this credit under LEED for New Construction and Schools. One point can be awarded for 5 percent of cost. A second point can be awarded for 10 percent of cost. Up to one point can be earned under LEED for Core & Shell for using at least 5 percent salvaged or recycled materials based on cost. An additional point can be earned as an Innovation in Design credit. This requires that the total value of salvaged or reused materials be more than 15 percent of the total materials cost.

The credit calculations for MR Credit 3 are based on either the replacement value for an equivalent new material, or the actual cost paid for the material—whichever is higher. Since some used products are repurposed in new projects, this is an important distinction. For instance, if old barn wood will be used as flooring in a new project, the replacement cost of comparable new flooring should be used in the calculation.

The primary materials considered under this credit include paving, foundations, site improvements, plantings, and building materials (*Figure 56*). Mechanical, electrical, and plumbing components are not eligible to be included. Appliances and equipment are also not eligible. Furnishings may be included if they are more than two years old. They may also be considered in subsequent material credits. Materials may only be included if they are permanently installed in the project.

To document this credit, the project team must keep a list of the material costs of the project. These costs exclude labor and equipment. It is important to ensure that material costs are accurately reported. This is especially true for materials procured by the Sustainable Construction Supervisor. Be especially vigilant if substitutions are proposed. These substitutions may be for salvaged, refurbished, or reused materials. Such requests should be carefully reviewed. It must be ensured that they do not negatively affect project sustainability goals under LEED.

3.4.2 MR Credit 4: Recycled Content Materials

MR Credit 4 encourages the use of materials with recycled content. The intent of this credit is to increase demand for recycled products and reduce the impacts associated with using new materials. *Figure 57* shows some examples of products with

201F56.EPS

Figure 56 Reused materials such as these wood I-joists at a Habitat for Humanity store are calculated by either actual cost or replacement value, whichever is higher.

recycled content that are used on building projects.

Projects under LEED for New Construction, Schools, and Core & Shell are eligible for up to two points under this credit. The credit is based on the quantity of recycled content products used in the project. An innovation credit is also possible for greatly exceeding the credit requirements under MR Credit 4.

The calculations for this credit require knowing both the pre-consumer and post-consumer percent recycled content for the product. Many products contain some of both types of recycled content. Therefore the calculation is based on the weight of each component with recycled content. It also includes the percentage of recycled content it has. After the composite pre- and post-consumer recycled content is determined for the product, the recycled content value is determined. This is calculated by multiplying these percentages times the total cost of the product or assembly.

The *LEED Reference Guide* contains more details on the calculations for recycled content credit. The Sustainable Construction Supervisor must know what information is required for the calculation (see checklist). As with MR Credit 3, mechanical, electrical, and plumbing materials are not included in calculations. Furnishings are only included if they are also included in all other relevant MR credits.

What's wrong with this picture?

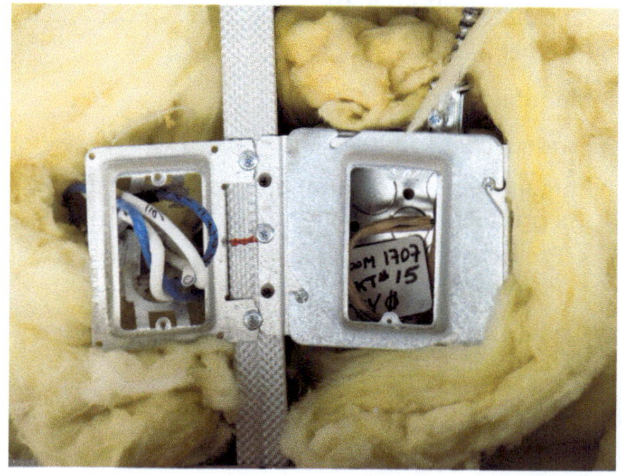

201SA05.EPS

A. Recycled plastic lumber
B. Insulating concrete form material with recycled polystyrene pellets
C. Recycled rubber turf stabilizer
D. Recycled paper fiber board
E. Recycled paper fiber countertop
F. Recycled rubber shingles
G. Recycled content carpet (face fiber and backing)
H. Salvaged lumber
I. Recycled paper and glass cement tiles
J. Recycled glass countertops
K. Recycled rubber flooring

201F57.EPS

Figure 57 Waste-based and recycled content materials.

CHECKLIST
Required Information for
MR Credit 4: Recycled Content

- ❑ Product name
- ❑ Manufacturer name
- ❑ Product cost
- ❑ Percent recycled content for each component of the product:
  - Pre-consumer recycled content
  - Post-consumer recycled content
- ❑ Percent of product weight for each component of the product
- ❑ For steel, the actual recycled content, if known
- ❑ For concrete:
  - Ratio (by weight) of supplementary cementitious materials to cement (flyash, ground granulated blast furnace slag, silica fume)
  - Dollar value of cementitious materials per unit of concrete
- ❑ Cutsheets or manufacturer's letters to document recycled content

MR Credit 4 is typically estimated during the design phase. To verify this credit, the project team must keep a list of the actual material costs of the project, excluding labor and equipment. It is important to ensure that material costs are accurately reported for this calculation. This is especially important for materials procured by the Sustainable Construction Supervisor.

Contractors and subcontractors should consult with vendors and suppliers to determine the availability of materials with recycled content. Recycled content should meet or exceed the project specifications. The contractor is responsible for making sure that specified materials are properly installed. The contractor is also responsible for tracking the cost of recycled materials used. Additionally, they must also track the quantity used. The contractor is also responsible for providing this documentation to the project team.

For products that contain only a single material or component, the weight of the material is not needed. Some products may have different percentages of recycled content for different components. For example, products such as carpet may have different backing and face fiber materials. For these products, the weight of each of these components per unit weight of the material is needed. Many manufacturers of products with recycled content can provide these calculations as part of product documentation.

Special assumptions and calculations are possible for both steel and supplementary cementitious materials. These include materials such as flyash or ground granulated **blast furnace slag**. The Sustainable Construction Supervisor should work closely with the project team to determine the best approach to account for these materials in the credit calculations. In particular, the Sustainable Construction Supervisor should contact the supplier to determine the actual recycled content of steel, if known. It is also important to contact concrete suppliers to determine the ratio of pre-consumer alternatives, such as flyash to other cementitious materials, in each mix design. The total dollar value of cementitious materials per yard of concrete is also required. This information can then be used to modify the recycled content calculations by the project sustainability team.

The Sustainable Construction Supervisor should be especially vigilant if substitutions for recycled content materials are proposed. Such requests should be carefully reviewed. It must be ensured that they do not negatively affect project sustainability goals under LEED. Any change in recycled content should be documented. This documentation should include using manufacturer literature for review by the project sustainability team.

3.4.3 MR Credit 5: Regional Materials

MR Credit 5 motivates the use of materials that have been extracted and manufactured locally or regionally. The intent of the credit is to reduce negative impacts from transporting materials long distances. The credit also contributes to supporting local economies. To fully qualify for this credit, building materials or products must have had all components harvested, extracted, or recovered within a 500-mile radius of the project site. The components must also have been assembled into the final product within this radius. Only products whose final assembly happened within a 500-mile radius of the project can be counted. If some part of the product comes from outside that 500-mile radius, only the fraction of the product from inside the radius may be counted (*Figure 58*). The percentage of product that counts is determined by weight. This is used to determine the cost that can be counted toward regional materials in the credit calculations.

Projects under LEED for New Construction, Schools, and Core & Shell are eligible for up to

201F58.EPS

Figure 58 Does this material meet regional material requirements? It depends on the location of the project site and the source for all of the components that make up the product.

CHECKLIST
Required Information for MR Credit 5: Regional Materials

- ❑ Product name
- ❑ Manufacturer name
- ❑ Product cost
- ❑ Location of harvest, extraction, or manufacture of each component of the product
- ❑ Location of final product assembly
- ❑ Distance between the project site and all product/component locations
- ❑ Percent of product weight for each component of the product
- ❑ Cutsheets or manufacturer's letters to document regionality

two points under this credit. This is based on the quantity of regional materials and products used in the project. An innovation credit is also possible for greatly exceeding the credit requirements under MR Credit 5.

The calculations for this credit require knowing the location of manufacture/assembly for the product and all of its components. Many building products contain a large number of components. The team should develop a strategy during preconstruction to focus on those products which are most likely to contribute to this credit. For instance, products that will comprise a large part of the total project cost are a good choice. These products are things such as concrete used in foundations, the structure, or parking areas.

Complex products with many components should be examined to determine whether it is worth the time to sort out their regionality. Some manufacturers of complex products have already determined locations where their products will meet the regional materials credit. The Sustainable Construction Supervisor may need to contact product manufacturers to find if this information exists.

The *LEED Reference Guide* contains more details on the calculations for the regional materials credit. From the standpoint of the Sustainable Construction Supervisor, it is important to know what information is required for the calculation (see checklist). As with MR Credit 3, mechanical, electrical, and plumbing materials are not included in calculations. Furnishings are only included if they are also included in all other relevant MR credits.

To verify MR Credit 5, the project team must keep a list of the actual material costs of the project. These costs do not include labor and equipment. Ensuring that material costs are accurately reported is important for this calculation. This is especially important for materials procured by the Sustainable Construction Supervisor.

Contractors and subcontractors should consult with vendors and suppliers to determine the availability of materials manufactured within the region. It is important when procuring materials for the project to be aware of products targeted to meet this credit. The contractor should be aware of which products have been targeted. If substitutions to those products are needed, the Sustainable Construction Supervisor should consult with the project sustainability team to ensure that the goals of the credit can still be met.

The contractor is responsible for ensuring that specified materials are properly installed. The contractor is also responsible for tracking the cost and quantity of regional materials used. The contractor must also provide this documentation to the project sustainability team.

The Sustainable Construction Supervisor should be especially careful with subcontractors. They may propose substitutions for regional materials. Such requests should be carefully reviewed. The substitution must not negatively affect project sustainability goals under LEED. Any change in regional materials should be documented. This documentation should include manufacturer literature for review by the project sustainability team.

3.4.4 MR Credit 6: Rapidly Renewable Materials

MR Credit 6 encourages the use of materials that can be rapidly renewed through biological processes. The intent of the credit is to reduce negative impacts from using materials that are non-renewable. It also considers materials that take a long time to be regrown. To fully qualify for this credit, building materials or products must consist of substances that can be sustainably harvested on a ten-year or shorter cycle (*Figure 59*). These products are typically plant-based. However, they can also come from animals. This is true in the case of sustainably harvested wool. For products containing multiple materials, the percentage of product that counts is determined by weight. Examples of these products are particleboard or furnishings. This is used to determine the cost that can be counted toward rapidly renewable materials in the credit calculations.

Projects under LEED for New Construction and LEED for Schools are eligible for one point under this credit. This credit is based on the quantity of **rapidly renewable materials** and products used in the project. An innovation credit is also possible for greatly exceeding the credit requirements under MR Credit 6.

Complex products with many components should be examined to determine whether it is worth the time to sort out their rapid renewability. Some manufacturers of complex products such as carpet have already prepared documentation and calculations to support this LEED credit. The Sustainable Construction Supervisor may need to contact product manufacturers to find out if this information exists.

The *LEED Reference Guide* contains more details on the calculations for the rapidly renewable materials credit. From the standpoint of the Sustainable Construction Supervisor, it is important to know what information is required for the calculation (see checklist). As with MR Credit 3, furnishings are only included if they are also included in all other relevant MR credits.

To verify MR Credit 6, the project team must keep a list of the actual material costs of the project. These costs do not include labor and equipment. Ensuring that material costs are accurately reported is important for this calculation. This is very important for materials procured by the Sustainable Construction Supervisor.

Contractors and subcontractors should consult with vendors and suppliers to determine the availability of materials with rapidly renewable

A. Soy-based foam insulation
B. Wool carpet
C. Corn-based carpet fiber
D. Coir/straw soil stabilization mats
E. Corn-based carpet backing
F. Coconut wood flooring
G. Wheat straw fiberboard
H. Sorghum board
I. Bamboo flooring/plywood
J. Linoleum
K. Hemp fiber fabric
L. Cork flooring
M. Cotton fiber batt insulation

201F59.EPS

Figure 59 Bio-based and rapidly renewable materials.

CHECKLIST
Required Information for
MR Credit 6: Rapidly Renewable Materials

- ❑ Product name
- ❑ Manufacturer name
- ❑ Product cost
- ❑ Percent by weight of product that meets rapidly renewable criteria
- ❑ Cutsheets or manufacturer's letters to document rapidly renewable content

content. When procuring materials for the project, the contractor should be aware of which products have been targeted to meet this credit by the project sustainability team. If substitutions to those products are needed, the Sustainable Construction Supervisor should consult with the project sustainability team to ensure that the goals of the credit can still be met.

The contractor is responsible for ensuring that specified materials are properly installed. The contractor is also responsible for tracking the cost and quantity of rapidly renewable materials used. The contractor must also provide this documentation to the project sustainability team.

The Sustainable Construction Supervisor should be especially careful with subcontractors. This is important if a subcontractor proposes substitutions for rapidly renewable materials. Such requests should be carefully reviewed. They must not negatively affect project sustainability goals under LEED. Any change in rapidly renewable materials should be documented. This documentation should include manufacturer literature for review by the project sustainability team.

3.4.5 MR Credit 7: Certified Wood

MR Credit 7 encourages the use of wood-based materials and products that have been certified as sustainably harvested in accordance with the Forest Stewardship Council's principles and criteria. The intent of the credit is to reduce negative impacts from conventional logging and timber harvest. It also seeks to reduce the damage to habitat. To fully qualify for this credit, wood-based materials or products must be certified by the Forest Stewardship Council (*Figure 60*). The calculation typically uses only wood products permanently installed in the project. Temporary products, such as formwork, bracing, scaffolding, temporary floor protection, and guardrails, may be included at the project team's discretion. However, if any temporary materials are included in the calculation, all such materials must be included. Temporary materials used on more than one project can only be counted toward one project's LEED calculations. Furnishings must be included if they have been consistently included in calculations for other MR credits.

201F60.EPS

Figure 60 A variety of certified wood products are available at popular distributors.

Projects under LEED for New Construction, Schools, and Core & Shell are eligible for one point under this credit. The credit is based on the percentage (by cost) of **certified wood** and wood-based products used in the project. The calculation compares the total material cost for certified wood products to the cost of all wood and wood-based products used in the project. An innovation credit is also possible for greatly exceeding the credit requirements under MR Credit 7.

The *LEED Reference Guide* contains details on the calculations for the certified wood credit. The Sustainable Construction Supervisor must know what information is required for the calculation (see checklist). For this credit, all new wood and wood-based products used on the project must be listed. Each product listing should identify which

CHECKLIST
Required Information for
MR Credit 7: Certified Wood

- ❑ Product name
- ❑ Manufacturer name
- ❑ Product cost
- ❑ Part or percentage of product that is FSC Certified (% FSC pure, % FSC mixed)
- ❑ Vendor chain of custody number and/or certificates
- ❑ Vendor invoices on a line-item basis showing chain of custody number
- ❑ Cutsheets or vendor letters to document FSC certification

components of the product are FSC Certified. The cost of products, such as windows, that contain both wood and non-wood components can also be counted but must be weighted to include only wood components.

To verify MR Credit 7, the project team must keep a list of the actual material costs of all new wood and wood-based products used on the project. These costs do not include labor and equipment. Ensuring that material costs are accurately reported is important for this calculation. This is especially true for materials procured by the Sustainable Construction Supervisor. The actual credit calculations depend on the proportion of pure versus mixed FSC Certified content and the cost of each component.

Contractors and subcontractors should consult with vendors and suppliers to determine the availability of materials with FSC Certified content. When procuring materials for the project, the contractor should be aware of which products have been targeted to meet this credit by the project sustainability team. If substitutions are needed, the Sustainable Construction Supervisor should consult with the project sustainability team to ensure that the goals of the credit can still be met.

The contractor is responsible for ensuring that specified materials are properly installed. The contractor is also responsible for tracking the cost and quantity of FSC Certified wood products used. The contractor must also provide this documentation to the project sustainability team.

The Sustainable Construction Supervisor should be especially careful if subcontractors propose substitutions for FSC Certified wood

Did You Know?

Some products are labeled as FSC Recycled or FSC Recycled Credit. These products do not count toward MR Credit 7: Certified wood. Instead, they count toward MR Credit 4: Recycled Content Materials.

materials. Such requests should be carefully reviewed. The substitution should not negatively affect project sustainability goals under LEED. Any proposed change in wood products or materials should be documented. This documentation should include manufacturer literature for review by the project sustainability team.

3.4.6 EQ Credit 4: Low-Emitting Materials

EQ Credit 4 rewards the use of interior building materials and products that comply with strict emissions limits established by third-party standards. The intent of the credit is to reduce the quantity of indoor air contaminants resulting from building products. These contaminants can be both irritating and harmful to the comfort and well being of installers and building occupants (*Figure 61*).

201F61.EPS

Figure 61 Low-emitting materials help preserve IAQ. This product label claims to meet or exceed all VOC requirements in all 50 states. It also lists specific chemicals not in the product.

Projects under LEED for New Construction and Schools, and Core & Shell are eligible for points under this credit based on whether the products used on the interior of the building comply with referenced standards. Interior products are defined for the purpose of this credit as being installed on-site inside the building's weatherproofing system. Products installed outside the weatherproofing system do not count toward the credit. *Table 3* shows the range of products covered under this credit.

This credit is often calculated as all-or-nothing. With this approach, if any products within a category do not comply, the project is not eligible for that point. The VOC budget method is another way to comply with credit requirements. This is especially important if use of some higher-VOC products is unavoidable. Using this method, products with lower-than-required VOC limits can be used to offset the higher-VOC products. This approach requires calculations to show that the actual VOCs offset during application remain within the limits established in the credit.

Different options exist for complying with some of the credit requirements. For instance, under EQ Credit 4.3, credit requirements are met if all flooring elements in the building's interior meet special test requirements established by the California Department of Health Services.

Additional credits are also available under LEED for Schools for furniture/furnishings and ceiling and wall systems. Some projects have received credit under LEED for New Construction Innovation in Design for low-emitting furniture and furnishings. For the furniture/furnishings credit (EQ Credit 4.5), furnishings must comply with GREENGUARD standards or State of California requirements. For the ceiling and wall systems credit (EQ Credit 4.6) all components must meet State of California requirements. This includes gypsum board, insulation, acoustical ceiling systems, and wall coverings. Additional information on these requirements is available in the *LEED Reference Guide*.

The *LEED Reference Guide* contains details on the reference standards and requirements to meet the low-emitting credit requirements. From the standpoint of the Sustainable Construction Supervisor, it is important to know what information is required to document compliance with

Table 3 Products Covered by EQ Credit 4: Low-Emitting Materials

Product	EQ Credit	Reference Standard
Adhesives, sealants, and sealant primers	4.1	South Coast Air Quality Management District (SCAQMD) Rule #1168
Aerosol adhesives	4.1	Green Seal Standard GS-36
Architectural paints and coatings	4.2	Green Seal Standard GS-11
Anti-corrosive and anti-rust paints	4.2	Green Seal Standard GC-03
Clear wood finishes, floor coatings, stains, primers, and shellacs	4.2	SCAQMD Rule #1113 for Architectural Coatings
Carpet	4.3	Carpet & Rug Institute's Green Label Plus requirements
Carpet cushion	4.3	Carpet & Rug Institute's Green Label requirements
Carpet adhesives	4.3	Must meet the requirements of EQ Credit 4.1, depending on adhesive type
Hard surface flooring, including vinyl, linoleum, laminate flooring, wood flooring, ceramic flooring, rubber flooring, and wall base	4.3	FloorScore standard requirements
Concrete, wood, bamboo, and cork floor finishes	4.3	SCAQMD Rule #1113
Tile setting adhesives and grout	4.3	SCAQMD Rule #1168
Composite wood and agrifiber products, including particleboard, medium density fiberboard (MDF), plywood, wheatboard, strawboard, panel substrates, and door cores	4.4	Must contain no added urea-formaldehyde resins or adhesives

credit requirements (see checklist). For this credit, all VOC-emitting products used on the project must be listed. Additionally, all of the VOC emission levels of the products must be listed. Keeping track of this information is especially important if a product with high VOC levels is used unintentionally. This is important because it may still be possible to prove compliance with credit requirements using the VOC budget method.

Contractors and subcontractors should consult with vendors and suppliers to determine the availability of products that meet reference standard requirements for low VOC emissions. When procuring materials for the project, the contractor should be aware of which products have been targeted to meet this credit by the project sustainability team. If substitutions to those products are needed, the Sustainable Construction Supervisor should consult with the project sustainability team to ensure that the goals of the credit can still be met.

The contractor is responsible for ensuring that subcontractors properly install specified materials. The contractor is also responsible for tracking which VOC-emitting products are used. They must also obtain information about those specific products. Finally, the contractor must provide this documentation to the project sustainability team. If the project sustainability team decides to use the VOC budget approach to achieve the credit, further documentation is required. The Sustainable Construction Supervisor must document the amount of product actually used during installation. This is much more time-consuming than avoiding VOC products altogether.

The Sustainable Construction Supervisor should be especially careful if subcontractors propose product substitutions for VOC-emitting products. Such requests should be carefully reviewed. These substitutions should not negatively affect project sustainability goals under LEED.

CHECKLIST
Required Information for
MR Credit 4: Low-Emitting Materials

- ❑ Product name
- ❑ Manufacturer name
- ❑ VOC emission levels, minus water, in grams per liter
- ❑ Allowable VOC limits from reference standard
- ❑ Amount of product used for the project (if using VOC budget)
- ❑ Cutsheets or vendor letters to document VOC levels of installed products

Any proposed change in products or materials should be documented. The documentation should include manufacturer literature for review by the project sustainability team.

3.5.0 Evaluating and Documenting Product Substitutions

Change is a regular and expected part of construction projects. If a project is seeking LEED certification, managing change becomes critical. Some material substitutions can disrupt the project's LEED goals. The Sustainable Construction Supervisor is responsible for reviewing and evaluating changes, and then taking actions to ensure that they do not cause problems for LEED certification.

3.5.1 Matching Products to LEED Credits

The first step in evaluating potential product substitutions is to determine how they might influence specific LEED credits. *Table 4* lists different types of changes that might happen on a construction project and the LEED credits that could be affected.

3.5.2 Determining Equivalence

After the potential impacts to LEED credits have been identified, the next step is to determine whether the proposed change will cause problems for the identified LEED credits. If potential problems may occur, reject the proposed change or bring it to the attention of the project sustainability team for verification.

Examples of proposed substitutions that could cause problems for LEED goals include the following:

- *New instead of salvaged materials* – This may mean that goals for MR Credit 3: Resource Reuse can no longer be met. New products may also cause air quality problems if they have synthetic components. Proposed substitutions should be checked for impact on recycled content, regionality, rapid renewability, and use of certified wood.
- *Leftover materials from other projects* – These substitutions could offer a benefit if the leftover materials can be counted as salvage. However, they may be difficult to document in terms of recycled content, regionality, rapid renewability, use of certified wood, and emissions. This could be a problem if the brand of material is unknown and labels/manufacturer literature is missing.

Table 4 Potential Product Substitutions versus Affected LEED Credits

Material Type	MR Credit 3: Materials Reuse	MR Credit 4: Recycled Content Materials	MR Credit 5: Regional Materials	MR Credit 6: Rapidly Renewable Materials	MR Credit 7: Certified Wood	EQ Credit 4: Low-Emitting Materials
Foundations/Substructure						
Concrete		X	X		X	
Steel	X	X	X			
Other...						
Superstructure						
Concrete		X	X			
Steel	X	X	X			
Wood	X	X	X	X	X	X
Other...						
Building Enclosure/Envelope						
Sheathing		X	X	X	X	X
Insulation		X	X	X		
Exterior finish		X	X		X	
Doors/windows		X	X	X	X	
Other...						
Interior Construction and Finishes						
Drywall		X	X			
Wood	X	X	X	X	X	X
Doors/windows		X			X	
Paint		X	X			X
Carpet	X	X	X	X		X
Tile/rigid flooring	X	X	X	X	X	X
Ceilings	X	X	X	X	X	
Other...						
Services						
Plumbing						
Electrical	X					
HVAC						
Fire suppression						
Other...						

- *Same material from a different supplier or manufacturer* – This could affect the regional materials credit calculation and should be checked with the project sustainability team or LEED Accredited Professional to determine whether it is critical or not for that project.
- *Change in product label or packaging* – Any time a product label changes in appearance, it is essential to verify that critical information remains the same. Depending on the credits for which a product is being counted, the new label or packaging should be checked to determine whether the product still complies.

3.5.3 Approving and Documenting Changes

As soon as the need for a product substitution is identified, review the proposed change and determine the potential implications for LEED credits and project sustainability. If there is a question regarding influence of the change on LEED credits, the Sustainable Construction Supervisor should consult with the project sustainability team before approving the change. Subcontractors should be made aware of the project's LEED goals. They should also be encouraged to consult with the supervisor to discuss proposed changes.

A product substitution may be proposed after the implementation of the project has already been planned. If this occurs, reviewing and documenting the change is critical. The following checklist can be used to assemble the necessary information to establish whether a proposed substitution is acceptable.

If changes are approved, addenda should be prepared to the LEED material calculation spreadsheets as necessary. The Sustainable Construction Supervisor may also keep a product change log. This log should document the details of the proposed change. This log should document who requested the change. It should also describe why the change was requested. A description of what the change entails should also be included. The log must also list which LEED credits may be impacted and indicate the final decision. These details can be useful in justifying decisions later in the project. Finally, documentation of changes to design-phase credits will be necessary during submittal of the construction-phase LEED documentation. The product substitution checklist will be useful in documenting the nature and outcome(s) of those changes.

CHECKLIST
Material Substitution Proposal Review

	Criterion	**Current Product:** ____________	**Proposed Substitute:** ____________
❑	MR Credit 3: Salvaged? (yes/no/NA)		
❑	MR Credit 4: Recycled content? • % pre-consumer • % post-consumer		
❑	MR Credit 5: Regionality? • Manufacture locations • Raw material locations		
❑	MR Credit 6: Rapidly renewable? (yes/no/NA)		
❑	MR Credit 7: Certified wood? (yes/no/NA) If yes, chain of custody number:		
❑	EQ Credit 4: Low-emitting material? • Adhesive/sealant • Flooring system • Composite wood/agrifiber • Furniture/furnishings • Ceiling/wall systems Allowable VOCs: ____________g/L		
	Quantity (unit)		
	Unit Cost		
	Total Product Cost		
	Acceptable Substitute?	❑ Yes ❑ No	
	Reviewed by:		
	Approved by:		

Review Questions

1. Using materials that reduce the amount of heat gained by a structure reduces _____.
   a. albedo
   b. heat island effect
   c. air contamination
   d. solar energy

2. Indoor finishes, such as carpet, furnishings, and ceiling tiles must be protected during installation because they may _____.
   a. be damaged during operation
   b. offgas harmful emissions if damaged
   c. impact proper air circulation in a building
   d. absorb substances from their environment

3. Using a product with recycled content _____.
   a. reduces the externalities of that product
   b. increases the service life of the product
   c. increases the virgin content of the product
   d. increases the embodied energy of the product

4. Using regional materials within a project _____.
   a. increases the amount of virgin material used in a project
   b. increases the recycled content
   c. decreases the likelihood of achieving LEED credits
   d. decreases the transportation required

5. A material is rapidly renewable if its components contain _____.
   a. pre-consumer recycled content
   b. post-consumer recycled content
   c. materials that can be rapidly regrown without damage to the environment
   d. materials that are both locally grown and locally harvested

6. In the BuildingGreen database, the information that is listed for each product is _____.
   a. its qualifications
   b. all of its certifications
   c. its recycled content
   d. its VOC content

7. A standard used by some companies in the construction industry to improve their environmental performance is _____.
   a. Energy Star
   b. ASTM International
   c. ISO 14000
   d. EPA

8. An independent, third-party certification and standards development agency that evaluates the environmental properties of building materials such as paint and housekeeping chemicals is _____.
   a. Forest Stewardship Council
   b. ISO
   c. ASTM International
   d. GreenSeal

9. To achieve an Energy Star label, a product must perform in the top _____ percent of all products within its class.
   a. 20
   b. 25
   c. 30
   d. 35

10. The WaterSense program applies to plumbing products in which sector of the construction industry?
    a. Landscaping and irrigation
    b. Residential and highway irrigation
    c. Residential and commercial
    d. Commercial and industrial

11. The GREENGUARD label indicates that a product meets strict _____.
    a. chemical emission standards
    b. particulate emission standards
    c. energy saving standards
    d. water saving standards

12. Products that have a chain-of-custody certification from initial harvest through final manufacture comply with which certification system?
    a. Green Label Plus
    b. GREENGUARD
    c. WaterSense
    d. Forest Stewardship Council

Review Questions

13. The organization best known for certifying plumbing system components such as piping and valves is _____.
    a. EnergyStar
    b. National Sanitation Foundation
    c. American Society of Testing and Materials
    d. WaterSense

14. The act of misleading consumers about the environmental benefits of a product or service is defined as _____.
    a. greenspeak
    b. greenwash
    c. greening
    d. greenwaving

15. The credit calculations for MR Credit 3 are based upon either the replacement value for an equivalent new material or the actual cost paid for the material, whichever is _____.
    a. higher
    b. lower
    c. more efficient
    d. available

16. The intent of MR Credit 4: Recycled Content Materials is to increase the demand for recycled products and _____.
    a. increase the percentage of pre-consumer recycled materials
    b. increase the use of virgin materials
    c. decrease the use of salvaged materials
    d. decrease the impact of using new materials

17. The calculation to determine the recycled content of materials is based upon the _____.
    a. weight of each component with recycled content
    b. volume of each component with recycled content
    c. transport distance of each component
    d. density of each component with recycled content

18. The intent of MR Credit 5: Regional Materials credit is to reduce _____.
    a. the amount of virgin material used
    b. the amount of post-consumer waste
    c. the amount of material required
    d. negative transportation impacts

19. Final tablulations of MR Credit 5: Regional Materials credit occur most often during which phase of the construction project?
    a. Pre-design
    b. Design
    c. Construction
    d. Post-occupancy

20. To determine the rapid renewability of a complex product with many components, the contractor should contact the _____.
    a. owner
    b. manufacturer
    c. designer
    d. vendor

21. Temporary materials such as formwork may be counted toward the FSC Certified wood credit for how many projects' calculations?
    a. Only one project
    b. All projects the wood is used for
    c. None of the projects
    d. Up to two projects

22. The calculation for MR Credit 7: Certified Wood typically only uses wood materials that are _____.
    a. temporarily used on the project
    b. permanently installed on the project
    c. grown and manufactured within the region
    d. used only in the occupied spaces of the project

23. To qualify for EQ Credit 4: Low-Emitting Materials Credit, a product must be installed _____.
    a. anywhere on the project
    b. inside the building's weatherproofing system
    c. outside the building's weatherproofing system
    d. in occupied spaces

4.0.0 Green Building Methods and Processes

The way in which products are assembled into buildings influences project sustainability. The goal of this section is to describe the role of the Sustainable Construction Supervisor in managing construction processes to improve sustainability. It also offers guidance for providing ongoing implementation support, verification, and documentation of LEED compliance for projects. This section focuses on process-related sustainable construction best practices. It also introduces important construction-phase LEED requirements, including indoor air quality management, waste management, materials storage, site disturbance management, erosion and sedimentation control, and verification of correct product installation. Specifically, this section provides the knowledge needed to perform the following tasks:

- Recognize discrepancies between construction documents and LEED requirements, and take action to avoid potential problems.
- Verify that compliant products and technologies are actually installed.
- Implement sedimentation/erosion control/**pollution prevention plans**, and document compliance with these plans.
- Implement site disturbance management requirements and document compliance with these requirements.
- Support the **commissioning** process throughout the construction phase.
- Implement construction waste management requirements and document compliance with these requirements.
- Implement indoor air quality management requirements and document compliance with these requirements.

Together, these actions establish the foundation of the Sustainable Construction Supervisor's contribution to project sustainability.

4.1.0 The Role of Construction Processes in Project Sustainability

Construction processes play a key role in project sustainability. This is true both during and after construction. Decisions about how to install products are as important as the choice of products themselves. This is especially true when considering the protection of workers on site, the surrounding sites and neighbors, future building occupants, and future generations.

4.1.1 Protecting Workers On Site

Choices of construction means and methods have immediate impacts for workers on the construction site. Chronic exposure to occupational hazards can cause serious long-term health problems for construction workers. These hazards include noise, dust, chemicals, and vibrations. Immediate job hazards, such as moving equipment, unstable earthwork, and working at heights, can also result in injury or death.

Green projects demand the utmost attention to worker safety and health. Some green products may reduce overall risks to workers. For example, low VOC paints and coatings will reduce exposure. However, workers may be tempted to avoid personal protective equipment if they believe the products pose no hazards. Proper training and work policies are essential to ensure that green products are installed safely, just like their conventional counterparts. Personal protective equipment and other engineering controls are just as much a requirement on green projects as on any conventional project (*Figure 62*).

4.1.2 Protecting Surrounding Sites and Neighbors

Green projects are not just about protecting the workers on the construction site. Green projects also require care for surrounding sites and neighbors. Construction can pose threats to neighboring sites from emissions in the air, water, and at soil level. In the air, threats to surrounding sites include noise and light pollution from construction operations; airborne dust, particles, and chemical emissions; and odors associated with construction operations or specific products.

201F62.EPS

Figure 62 Proper protective equipment and controls ensure worker safety.

What's wrong with this picture?

201SA06.EPS

At the level of water and soil, contaminants can result from stormwater runoff and erosion, groundwater contamination, and soil contamination. These problems are mostly an issue during construction. However, they can also continue during building operations as well.

Green projects involve active measures to prevent or mitigate these threats. The law requires careful attention to stormwater management, erosion, and sedimentation control (*Figure 63*). Additional pollution prevention measures are encouraged by LEED. These include the reduction of air and light pollution during construction. Green buildings often include stormwater management measures that will continue to prevent pollution of surrounding sites after construction is complete.

4.1.3 Protecting Future Building Occupants

A third aim of green projects is to protect future building occupants. Measures during construction that help to achieve this aim include the following:

- Use of low-emitting products and materials
- Proper housekeeping practices during construction

Figure 63 Erosion control measures protect neighboring sites from sedimentation.

- Isolation of polluting activities from HVAC and occupied building areas (*Figure 64*)
- Sequencing of construction activities to avoid introducing contaminants into building finishes or finished areas
- Proper material storage to protect vulnerable materials from moisture damage that can lead to mold
- Control of the interior environment during later stages of construction to prevent excess humidity and mold growth
- Conducting a building **flush out** before occupancy

Each of these measures involves reducing the likelihood that future building occupants will be exposed to harmful chemicals, particulates, and biological contaminants from the construction process.

201F64.EPS

Figure 64 Protecting HVAC system components during construction prevents future air quality problems.

4.1.4 Protecting Future Generations

Finally, building sustainability is ultimately about protecting future generations and their ability to enjoy the same quality of life we have. This goal can be realized by choosing construction processes that:

- Avoid, mitigate, or restore damage to natural ecosystems
- Minimize the use of energy and natural resources
- Recover usable or recyclable materials for use in other projects (*Figure 65*)
- Result in healthy buildings without future environmental liabilities
- Result in buildings that are adaptable and efficient to operate and maintain

These objectives guide the selection of construction processes for green projects.

4.2.0 Sustainable Construction Operations

Managing a project to increase its sustainability requires careful attention to construction operations. Minimizing disturbance to the construction site helps preserve habitat and **biodiversity** that may exist on the site. It also makes it easier to restore the site when the project is complete. Careful control of erosion and sedimentation retains valuable soil on site. It also prevents damage to surrounding sites and waterways. Measures to prevent light, noise, and dust pollution during and following construction also preserve a good relationship with surrounding sites. Many opportunities exist for environmental stewardship within the project office. There are also opportunities in traveling to and from the project site. Finally, all of these measures combined can be part of a sustainable operations and pollution prevention plan for the project.

4.2.1 Minimizing Site Disturbance

One of the most important strategies for sustainable site operations is to minimize disturbance to the project site during construction. Except for **brownfield sites** that require environmental remediation, the less disturbed the construction site, the better. Particularly important are greenfield (previously undisturbed or undeveloped) sites. Also important are sites with critical habitat or important ecological features. However, minimizing disturbance even on previously developed sites creates fewer problems that must be fixed at the end of the project. It also reduces environmental and legal liability during construction.

Disturbing the site creates potential soil erosion and instability problems. It can also damage tree roots, even if the tree trunk and branches appear to be undisturbed. Careful attention must be paid to limiting or prohibiting vehicle traffic in areas with tree roots. This is especially true for heavy equipment. The weight of the vehicles and equipment cause **soil compaction** and damage the ability of the tree roots to absorb nutrients from the soil. This may eventually kill the tree.

The most common means for limiting site disturbance is to erect construction fencing as a boundary for operations. Chain link or security fencing is often used around the perimeter of the site for security purposes. However, lightweight plastic fence is more often used within the site

201F65.EPS

Figure 65 Material recycling conserves resources for future generations.

What's wrong with this picture?

201SA07.EPS

boundaries to mark areas that should be left undisturbed. As shown in *Figure 66*, bright orange plastic fencing is erected using stakes to provide a clear visual boundary for construction equipment. The fence is provided as tree protection fencing for the mature tree to its left. The fence is placed at least as far from the trunk as the drip line of the widest branch. This ensures that heavy equipment will be restricted to areas away from the tree roots.

Construction fencing can and should be used to isolate critical limits of disturbance throughout the site. A second important factor in limiting site disturbance is proper selection of construction equipment. Equipment is often selected based on minimizing cost to the project. Since labor has historically been the most expensive aspect of construction projects, equipment is often selected to maximize human productivity and minimize the length of time a task requires.

This approach, however, does not take into account the ecological damage heavy equipment can cause. For example, metal-tracked equipment can cause serious damage to site soil. Rubber tracked equipment (*Figure 67*), and tire-based equipment may cause less damage depending on soil conditions. While ecological damage may not be the only criterion for equipment selection, it should at least be considered as a factor in making equipment choices.

Contractual penalties and incentives can also be used to motivate more sustainable behavior on site. For instance, projects for which preservation of mature trees is important may impose fines for damaging trees on site. This fine is often a function of tree diameter. One project manager in Atlanta, GA posted large price tags on the trees to be preserved on his project site (*Figure 68*). This made the contractual penalties for damaging trees very clear to the excavation subcontractor and its equipment operators. As a result, no trees were damaged during the project.

FENCING TO LIMIT DISTURBED AREA

TREE PROTECTION FENCING

201F66.EPS

Figure 66 Fencing can be used to limit disturbed areas (top). Tree protection fencing should be at least as far from the trunk as the drip line of the outermost branches (bottom).

201F67.EPS

Figure 67 Metal tracked equipment does more damage than rubber tracks on soft ground.

201F68.EPS

Figure 68 Signs remind equipment operators of contract penalties for damaging vegetation.

201F69.EPS

Figure 69 Silt fencing is a common strategy for erosion and sedimentation control.

4.2.2 Erosion and Sedimentation Control

During project planning, site selection, and design, the project sustainability team can take many actions that prevent erosion and sedimentation during construction. These include choosing previously developed sites instead of greenfield sites, choosing sites with few steep slopes, and locating buildings on areas of the site that require less grading.

Managing erosion and sedimentation is the responsibility of the general contractor and subcontractors. Ensuring proper control of erosion and sedimentation is a fundamental responsibility of the Sustainable Construction Supervisor (*Figure 69*). All projects larger than one acre or part of a development that is larger than one acre are required by law to meet **National Pollutant Discharge Elimination System (NPDES)** requirements. Projects seeking certification under the LEED rating system must also meet these requirements, even if they are smaller than one acre in size.

The project civil engineer or landscape architect typically begins the control process by identifying erosion-prone areas of the site. They then outline soil stabilization measures (see *Table 5*). However, the contractor must develop an Erosion and Sedimentation Control (ESC) plan. This plan must implement those measures and meet NPDES requirements. The Sustainable Construction Supervisor is then responsible to ensure that the plan requirements are met. This is especially important immediately following storm events (*Figure 70*).

The best measure for erosion and sedimentation control is to avoid disturbing site vegetation during construction. This can be accomplished through careful planning of site operations to minimize the areas of the site that will be disturbed.

In areas where site disturbance is unavoidable, construction activities should be carefully planned to minimize potential damage. Areas where disturbance will be frequent should be isolated from the rest of the site by structural controls such as earth dikes, wattles, or silt fences. Stormwater falling onto these areas should be diverted to **sediment** traps or sediment basins to allow soil particles to settle out. Mulching with gravel can be used to stabilize areas where equipment or vehicle traffic will occur. Gravel can also help control airborne dust.

Temporary or permanent grading can help to slow and divert stormwater to desired areas. Where possible, grading should be done to encourage sheet flow instead of concentrated flow of stormwater. Sheet flow involves water running at a low depth across a wide area. Sheet flow tends to slow water velocity, thereby reducing erosion potential. Lengthening flow paths and reducing gradients also help to reduce water velocity and minimize erosion.

Table 5 Erosion and Sedimentation Control Measures

Control Measure	Description
Soil Stabilization	
Temporary Seeding	Plant fast-growing grasses or "green manure" species to temporarily stabilize soil in areas where the soil will again be disturbed in the future.
Permanent Seeding	After soil has been placed in its final location on site, plant grass, trees, shrubs, and other plantings to permanently stabilize soil.
Mulching	Place straw, cut grass, wood chips, or gravel on the surface of disturbed soil to hold it in place. Choose an organic mulch that will quickly decompose if the soil will later be moved to another location.
Structural Control of Soils	
Earth Dike	Construct a dam of stabilized soil to divert runoff from disturbed areas into sediment basins or sediment traps. Use seeding or other measures to stabilize the earth dike.
Wattle	Install bundles of straw or straw bales continuously across areas of erosion to remove sediment from stormwater before it leaves the site. Wattles should be staked in place to prevent movement.
Silt Fence	Install 2" × 2" stakes no greater than 6' apart to support filter fabric to remove sediment from stormwater before it leaves the site. The base of the filter fabric should be placed in a trench at least 4" × 4" and backfilled with native soil or gravel to prevent blow-out.
Sediment Trap	Excavate a pond area or construct earth embankments to contain stormwater. This trap will retain stormwater and allow sediment to settle while the water percolates into the soil.
Sediment Basin	Construct a pond with a controlled water release structure to allow sediment to settle out of stormwater before it leaves the site.

201F70.EPS

Figure 70 Unprotected soil becomes unstable during storm events.

What's wrong with this picture?

201SA08.EPS

"Understanding of project goals and communication among all involved are the keys to success for any project—green or otherwise. Education regarding sustainable alternatives allows those involved to see both the big picture intent and details needed to meet the project requirements."

James G. Sullivan, Ph.D, LEED AP BD & C, Assistant Professor, M.E. Rinker, Sr. School of Building Construction, University of Florida

Diverting stormwater into vegetated areas can slow water velocity. It also encourages sediment filtering and water percolation. Sediment basins can and should be vegetated. This vegetation can be temporary or permanent seeding to promote filtration and absorption of stormwater. More permanent vegetated structures are likely to be part of sustainable site design for green projects. Examples of these structures are **infiltration trenches**, **vegetated swales**, and **bioretention cells**. The general contractor should work closely with the project design team to determine the potential for incorporating such structures. At a minimum, grading for them as part of stormwater management during construction should be considered.

Areas with infrequent disturbance, such as spoil piles or soil storage areas, may rely upon temporary seeding or mulching to minimize damage to the soil before it is replaced on site. When choosing a seeding strategy, use native or non-invasive plants to minimize the potential for future landscaping problems. Some plant species are known as green manure. These species become quickly established after planting and fix nitrogen in the soil in which they are grown. They can be tilled under and mixed into the soil during grading. When this happens, they enrich the soil as compost that helps permanent landscaping become established. Recommended species of green manure that work well for this purpose include legumes (members of the pea family) and various types of clover. Consult a local agricultural school or the county extension office for recommended species for the local climate and conditions.

4.2.3 Pollution Prevention

After construction begins, pollution of many types can threaten workers on site and neighboring sites. Pollution can be defined as any harmful substance or effect introduced into the environment as a byproduct of another activity. Noise, dust, air pollution, and engine drippings from heavy equipment all reduce the sustainability of construction operations (*Figure 71*). Emissions and offgassing from materials can contaminate the air. Accidental spills or improper disposal of wastes contaminates the soil. Light pollution during night operations can pollute the environment if it shines beyond the work area. Sediment and eroded soil can pollute neighboring sites if allowed to run off the project site.

Reactive measures can and should be taken on site to protect workers from these threats. Personal

201F71.EPS

Figure 71 Equipment running over dry soil generates airborne dust.

What's wrong with this picture?

201SA09.EPS

protective equipment (PPE) is one way to ensure that workers are protected from hazards such as noise, vibration, chemical exposure, and struck-by injuries. Likewise, reactive measures, such as perimeter silt fences, can be taken to stop sediment before it runs off the site.

However, reactive measures assume that pollution problems are going to happen in the first place. A better approach is to use solutions that prevent problems before they happen. *Table 6* shows examples of proactive versus reactive solutions to common types of construction pollution.

Table 6 Examples of Pollution Prevention and Mitigation Measures

Pollutant	Source(s)	Reactive (Mitigation) Measures	Proactive (Prevention) Measures
Light	Night operations Welding or cutting operations Temporary lights left on at night	Shield or redirect light fixtures to focus only on work site Turn off temporary lights at end of work day	Revising construction schedules to avoid night operations Using smaller lights focused directly on task areas
Noise and vibration	Equipment operation	PPE for workers Arrange work shifts to allow worker breaks Perimeter fencing for noise barrier	Revising construction schedule to avoid operations during sensitive times Choosing equipment with lower noise production
Dust and airborne particulates	Equipment operation Wind erosion of exposed soils	PPE for workers Surface treatment of exposed soils with water or dust suppression chemicals	Limiting sit disturbance Leaving existing vegetation intact Covering exposed soil with temporary or permanent seeding
Airborne chemical emissions	VOCs from the offgassing of new synthetic materials	PPE for workers Increase ventilation rates during product installation	Using low- or no-VOC products Designing for exposed surfaces Using prefinished materials
Soil and groundwater pollution	Engine drippings Refueling Accidental spills Improper disposal	Spill countermeasures such as berms, absorbent mats, and barriers Provide contained storage for chemicals and hazardous materials Spill cleanup plans/equipment	Using non-hazardous materials where possible Spill prevention training for employees Proper equipment maintenance Centralized refueling
Surface water pollution (heat and contaminants)	Engine drippings Accidental spills Exposed soil without erosion control measures Paved surfaces	Spill countermeasures Perimeter silt fences Spill cleanup plans/equipment Provide contained storage Stormwater detention basins	Proper equipment maintenance Pervious or high-albedo surfaces Seeding exposed soil Limiting construction disturbance Infiltration basins
Tracked soil on neighboring streets	Vehicle wheels	Vehicle wash stations	Limiting construction disturbance Off-site materials staging Just-in-time delivery

What's wrong with this picture?

201SA10.EPS

201F72.EPS

Figure 72 Recycling facilities help to divert waste in site trailers.

4.2.4 Site Business Operations

In addition to the construction activities used to create the building, other opportunities exist to be sustainable. The contractor business operations on site can contribute to project sustainability. The same practices that will be employed in the final building should also be considered for the site's temporary construction office. For instance, recycling facilities for paper, beverage containers, and other types of recyclable waste should be available for workers on site (*Figure 72*). Other containers may be appropriate to collect rechargeable batteries, compostable food waste, or other types of waste.

Closing the recycling loop is also important. Using recycled content paper and office products helps to provide a market for recyclable materials. Refurbished office furniture also saves raw materials and reduces waste. Paper goods, such as toilet paper, can be obtained with recycled content. Recycled content paper towels can be collected and composted if facilities are available.

Nearly every aspect of site operations has opportunities to improve environmental stewardship. For example, it is possible to rent environmentally friendly portable toilets in some markets. These toilets use non-toxic chemicals that are formaldehyde- and alcohol-free. During servicing, they may be cleaned with biodegradable, non-toxic cleaning products. Paper products may be chlorine-free and made from recycled content. Water used in flushable stalls can also be dispensed through water-conserving fixtures. A web search for green portable toilets will identify what options are available for a particular job site.

Rental of temporary construction barriers and fencing is another way to reduce job site impacts. This reduces the use of virgin materials to construct the fence. It also reduces the need to dispose of temporary materials at the end of the job. Similarly, with the advent of Building Information Modeling (BIM), job sites may become increasingly paperless in the future. Plans, specifications, and other project documents may be shared among the project team electronically rather than through paper copies. The LEED Online system for managing LEED project submittals is one example of this type of approach. It reduces the need for raw materials on which to print documents. It also reduces the energy needed to print them. It also reduces the need for physical storage space at the end of the project.

Virtual project meetings are now possible. These meetings can use the widely available, low cost infrastructure available on many projects. Conference calls and video or computer-based conferencing can replace the need to physically travel to the job site. Webcams, remote sensing, and other digital technology can provide documentation of project conditions. These also reduce the need for physical transportation to the site.

Site utilities, such as power and water, should be carefully managed as well. Temporary utility lines are often constructed without the same level of care as permanent lines. This can result in water leaks. This is particularly true if flow is not shut off after the water has been used. Be aware of the location of all temporary utility lines and regularly inspect them to locate and repair leaks. Instruct subcontractors to immediately bring leaks to the attention of the supervisor. Turning off unneeded lights and other equipment when the site is unoccupied should also be emphasized to subcontractors. These methods provide another way to make the project more environmentally friendly and resource efficient.

Transportation of people and materials to and from the site also has impacts. These impacts are on resource efficiency, air quality, and traffic congestion. Especially for congested sites, Sustainable Construction Supervisors should consider the possible impacts of employee parking on site preservation goals. Policies and incentives to encourage employee carpooling or transit use can help to reduce these problems. Consider providing preferential parking for carpools and alternative fuel vehicles when designing the site layout.

Some firms have converted to alternative fuel vehicles for their fleets (*Figure 73*). Employees may also elect to drive their own alternative fuel vehicles to the project. Increasingly, heavy equipment operates on biodiesel mixtures instead of conventional diesel. This reduces the air emissions resulting from diesel combustion and also reduces dependence on non-renewable resources.

Managing site access can also have a significant environmental impact. Sites with security inspection checkpoints can be particularly problematic. One example on an Air Force base in Oklahoma actually involved moving the whole project site outside the security fence for the duration of construction. This eliminated the need for long periods of vehicle idling while waiting at the security checkpoint. It also reduced the risk of potential problems with products. An example is ready-mix concrete, which has a limited time for placement after the materials have been mixed. This project received a LEED innovation credit for reducing the overall carbon footprint of construction with this simple but effective measure.

Careful planning of material deliveries and equipment mobilization and demobilization can reduce air quality impacts of vehicle idling. Reducing transportation impacts is an important part of a sustainable site operations plan.

201F73.EPS

Figure 73 Alternative fuel vehicles, such as this hydrogen car, reduce carbon footprint.

4.2.5 Developing a Sustainable Operations and Pollution Prevention Plan

Sustainable Construction Supervisors should approach site operations and pollution from a proactive, not reactive, stance whenever possible. In this way, potential problems can be averted. Reactive measures are still available if needed. Careful thought about the potential sources of pollution during project planning can help to reduce problems later. *Table 7* provides a checklist of construction activities likely to result in pollution. Each of these activities should be carefully reviewed to identify potential pollution specific to the project and then used to develop a pollution prevention plan. Identify practices that can be used to prevent or mitigate potential problems.

Opportunities to reduce resource consumption, waste generation, and other harmful impacts should also be inventoried at the start of the project and used to develop a pollution prevention plan. These include office practices, project documentation practices, green services and utilities, and transportation.

Sustainable site operations plans should include an inventory of all of these potential impacts. The

Table 7 Potentially Polluting Construction Activity Checklist

Construction Activity	Sediment/ Dust	Nutrients	Trace Metals	Pesticides	Oil, Grease, Fuels	Other Toxic Chemicals	Light/ Noise
Construction Practices							
Dewatering operations	X					X	X
Paving operations	X			X	X	X	X
Earthwork	X				X		
Landscaping	X	X		X			
Structure construction/ painting			X			X	X
Material Management							
Material delivery/storage	X	X	X	X	X	X	
Material use		X	X	X	X	X	
Waste Management							
Solid waste	X	X					
Hazardous waste						X	
Contaminated spills	X					X	
Vehicle/Equipment Management							
Vehicle/equipment fueling						X	
Vehicle/equipment maintenance						X	

What's wrong with this picture?

201SA11.EPS

checklist for Sustainable Site Operations lists impacts that should be considered as part of the plan. After the impacts have been identified, applicable best practices should be identified that can reduce or eliminate them. Responsibility for monitoring and quality control may be assumed by the Sustainable Construction Supervisor or assigned to others. Responsible parties should be clearly indicated as part of the sustainable site operations plans.

4.2.6 Sustainable Operations and Pollution Prevention Plan Implementation

Keys to the success of a sustainable site operations and pollution prevention plan are the knowledge and motivation to make the right choices on the job site. Preventing pollution and minimizing negative impacts of site operations requires participation by everyone on site. Important considerations include the following:

- *Alignment with corporate values* – Sustainable operations will be most successful when they are part of a larger corporate philosophy, not just temporarily used for a special green project. Contractor employees should be familiar with the company's position on sustainability. Supervisors should undergo sustainable construction training to be able to provide guidance to employees on how to behave. Corporate support for recycling, procuring recycled content materials, and alternative fuel vehicles will help to ensure that these tactics are used uniformly on all projects. Including sustainability as part of performance evaluation and promotion criteria further validates the importance of the concept to employees.

CHECKLIST
Sustainable Site Operations – Impacts and Opportunities

Site Office Practices

- ❑ Office waste recycling (paper, cardboard, containers, etc.)
- ❑ Operations waste recycling (rechargeable batteries, compostable food waste, etc.)
- ❑ Recycled-content paper and office products
- ❑ Composting paper waste (paper towels)

Project Documentation Practices

- ❑ Electronic documentation, including photos and/or webcam
- ❑ Building information modeling
- ❑ LEED Online
- ❑ Virtual meetings

Green Services and Utilities

- ❑ Portable toilets
- ❑ Temporary construction barriers
- ❑ Maintenance and inspection of temporary water lines
- ❑ Policies and incentives for energy conservation
- ❑ Layout and controllability of temporary lighting

Transportation

- ❑ Preferred parking for carpools and alternative fuel vehicles
- ❑ Alternative fuel fleet vehicles
- ❑ Carpool policy and incentives
- ❑ Transit policy and incentives
- ❑ Site access control and congestion
- ❑ Impacts of on-site parking areas
- ❑ Reduction in construction equipment lighting

Education and Incentives

- ❑ Employee training/orientation
- ❑ Subcontractor training/orientation
- ❑ Signage and posted information
- ❑ Policies to encourage environmentally responsible behavior

- *Employee training and orientation* – Basic sustainability training and new employee sustainability orientation are an important part of introducing employees to the corporate sustainability policy. Education is also essential to ensure employees know what to do when on the job. Education will help general contractor employees answer subcontractor questions that may arise.
- *Subcontractor training and orientation* – Providing training or basic orientation for subcontractors on a job will help them comply with requirements to meet project sustainability goals. Regular reminders of proper behavior can be provided at toolbox meetings or during informal project walk-throughs.
- *Signage and posted information* – Proper signage or posted information will help employees and subcontractors make correct choices for project sustainability goals. Information at the proper level of detail is especially important at points on the job site where decisions are made. These include recycling dumpsters, material stockpiles and cutting stations, and near utility cutoffs. Information may need to be provided in more than one location, depending on the composition of the work force.
- *Celebrating success* – Modifying behavior is most effective when feedback is provided. Employees and subcontractors should be kept informed of progress toward sustainability goals. Measures, such as weekly **waste diversion** rates, can help employees gauge the effectiveness of their efforts. Information should be followed up with opportunities for discussion and further education on how to improve performance. It may also be linked with performance incentives, recognition, or other rewards.

4.3.0 Materials Management

Effective materials management improves project sustainability. It also offers the potential to save project costs. Working with vendors on product procurement and delivery practices can reduce solid waste and increase efficiency. Better product storage helps prevent damage to products. It also saves the cost of later replacement or disposal of a damaged product. Finding alternative things to do with excess materials reduces disposal costs and may also offer benefits in terms of tax credits or vendor credit. All of these tactics can be integrated into a **materials management plan** that will save money while improving project sustainability.

4.3.1 Product Procurement/Delivery

Not only is it important to choose green products for use in a sustainable construction project, it is also important how those products arrive on site. Two key elements of this process should be considered on green products: product procurement and product delivery.

Product procurement involves identifying and selecting a source for products. It also involves communicating product requirements and delivery expectations to that source. It is followed up by ensuring that what is delivered meets those requirements. Additionally, it involves working with the vendor or supplier to correct any problems. Procuring green products may require using different vendors and suppliers than is customary for a company. New relationships, accounts, and lines of communication may need to be established.

These relationships are a good investment for companies working on green projects. However, they represent a risk of the unknown. Lead times for products may be unfamiliar. More time may need to be built into the schedule. New requirements for product documentation may also require additional attention. Manufacturer training and certification may also be required for some types of products.

Building these relationships early on will help companies be more competitive as green projects become the norm. Obtaining necessary training and building a product documentation file for green products will reduce the learning curve and uncertainty for future projects. Systematic attention to product procurement is worth the effort in the long run.

The second half of the equation is product delivery. An important part of green delivery is the means and methods of transportation used to deliver the product to the site. This may be difficult to influence. A second important consideration is the packaging that is used to transport and protect the product during delivery.

Adequate packaging is essential to prevent damage to products during and following delivery. However, that packaging need not be an environmental liability. Some manufacturers can provide their products with returnable or reusable packaging (*Figure 74*). This prevents the packaging from turning into waste assuming that it is properly returned to the source. For instance, delivery of small sizes of sand and aggregate can be arranged using returnable heavy-duty bags that are removed from the delivery truck by crane. These bags allow multiple types of materials to be delivered at once. This has lower transportation impacts than bringing loose material in a truck bed. It also keeps these materials from being contaminated on site or spreading to unwanted areas, thus speeding up clean-up activities.

Other types of packaging may be compostable or biodegradable. For instance, many types of plastic are now being made from plant products, such as corn and soybeans. Depending on environmental conditions, these plastics may be compostable, especially in municipal **composting** facilities where higher temperatures are maintained than in home composting systems. Be sure to check with local composting facilities to verify acceptability of these types of wastes.

Still other construction material packaging can be recycled. Corrugated cardboard is a fairly high value waste for recycling and can be recycled in most locations. Storing it in a protected location helps keep the material dry. It also makes it easier to handle for recycling. Wood pallets that are not reusable can be recycled along with clean wood waste. They can also be chipped and used as mulch.

What's wrong with this picture?

201SA12.EPS

201F74.EPS

Figure 74 Reusable packaging reduces construction waste. This pallet is reusable, and the plastic covering the load may be compostable.

Molded polystyrene is commonly used with cardboard to secure products within packaging. It is difficult to recycle in all but a few locations. Sheet polystyrene or packing peanuts may be reusable. Investigate local shipping outlets to see if they accept such materials for reuse. Waste paper can be used, either shredded or crumpled, as packaging material that may then be recycled or composted. Packing peanuts may also be made of rapidly renewable, starch-based biodegradable materials. Be sure to verify that this is the case before mixing these items with compostable waste. Sometimes popcorn is even used as an environmentally friendly packing material!

After packaged material arrives on site, it must be stored in a safe, secure location until it is needed. One way to minimize both the need to store material and the risk of damage to the product is **just-in-time delivery**. Just-in-time delivery is fairly common for projects involving large components for which on-site storage would be difficult (*Figure 75*). For such projects, it requires considerable planning and coordination since the components are likely to be custom-fabricated.

Just-in-time delivery can also be used for situations where materials are more readily available. It can also be used when scheduling constraints are looser. Certain commodities, such as drywall, drop ceiling panels, carpet, carpet pad, and insulation, are best delivered as close to installation as possible. These materials are highly absorptive. They pose a high risk of moisture damage and can also be easily damaged while in storage. They also can easily absorb contaminants from other parts of the project. These contaminants can then escape into the building after the materials are installed.

201F75.EPS

Figure 75 Just-in-time delivery and placement of products such as precast concrete reduces the need for space on site.

4.3.2 Product Storage and Staging

The construction schedule dictates when various materials will be needed for a project. Depending on how predictable and reliable the supply chain is, having products arrive on site well before they are needed may be necessary to avoid construction delays. When products arrive before they can be installed, adequate product storage and staging must be provided to ensure their protection.

Material security is one possible threat to stored materials. Other threats result from environmental conditions, such as moisture or temperature changes. Still other damage, such as crushing or puncturing, can be caused during construction operations. Other problems may occur from chemical spills or absorption of contaminants from the surrounding environment.

Preventing moisture damage is of paramount importance in material storage. This is especially important for products that are absorptive. Exposure to moisture increases the likelihood of mold in the product itself. It can also increase the likelihood of mold in adjacent products after installation. Moisture can also damage adhesives used in composite materials. It can also cause swelling of particles. Careful attention to moisture from all sources is worth the effort. This includes absorption from the ground and the air. Stacking materials loosely to allow good air circulation is also advised in most situations. In extremely humid climates, however, obtain manufacturer advice on how best to store materials. Be certain to consider just-in-time delivery where possible.

Other potential damage can be caused by exposure to ultraviolet (UV) radiation. Products containing plastics, especially, must be protected

What's wrong with this picture?

201SA13.EPS

from UV exposure or they begin to photodegrade. Rigid plastics become more brittle. Softer plastics may become chalky or otherwise lose their integrity. Color-critical products may start to fade. Covering or shading materials to prevent UV exposure is a critical part of sustainable material storage.

Preventing outside damage is not the only action necessary for material storage. Materials themselves may require attention prior to installation. For instance, materials containing synthetic components or adhesives may need to be unwrapped and allowed to offgas before installation. This prevents potential indoor air quality problems after the building is complete. Products containing fabrics, foam, composite wood, adhesives, and finish materials all may need to offgas (*Figure 76*). During this time, they must also be carefully protected from other types of damage.

4.3.3 Excess Materials

After products have arrived on site and have been installed in the building, any remaining excess materials must be addressed. Some materials may remain with the building as attic stock for future repairs or maintenance. Others may not be suitable for subsequent use and must be disposed of.

However, some materials may remain in smaller quantities that could still be used for other purposes. Finding good uses for these products is a critical part of both materials management and effective waste management.

Ideally, leftover materials should decrease if proper storage and staging are done. They will also decrease if careful estimating and procurement are done. Sustainable construction projects should prevent pollution by not ordering more material than necessary to get the job done. However, this may not always be possible if getting replacement materials would be difficult and time-consuming. This is especially true for unusual or hard-to-get products.

Leftover or excess materials can sometimes be returned to the vendor, supplier, or manufacturer for reuse. It is important to protect excess materials from potential damage (especially moisture and UV damage) to enable product restocking.

In other cases, usable leftover materials can be donated to charities, such as Habitat for Humanity. Local universities, colleges, or trade schools may also be interested in leftovers to use as part of education and training, particularly if they are innovative new products. If no other options are practical, excess materials can be given away through posting on listservs, such as www.freecycle.org. Interested parties who see a post on Freecycle contact the poster and arrange to pick up the materials. They may even help remove used materials from the building during demolition. This reduces labor requirements and ensures that products are removed in a way that preserves future use. When posting on Freecycle, be sure to read and follow appropriate posting etiquette. Email access is required in order to use this system. Even very small quantities of new or used materials may be useful to others. Don't discount the possibility.

201F76.EPS

Figure 76 Furnishings and cabinetry should be allowed to offgas away from the building prior to installation.

4.3.4 Developing a Materials Management Plan

As with other types of management plans, developing a materials management plan requires an inventory of needs. It also requires the identification of potential solutions. Finally, it requires a cost/benefit analysis to determine the best course of action. Needs for materials management will correlate closely with the construction schedule. Therefore, the construction schedule is a good starting point for identifying needs and requirements.

A materials management needs inventory should begin by identifying materials associated with each construction task. Then an evaluation of the procurement and storage/staging requirements for each type of material should be done. *Table 8* lists typical information that should be compiled for each type of material. It should be prepared in consultation with subcontractors. This will ensure that their material storage/staging needs are also met.

After all material storage needs have been inventoried, the next step is to map these needs over time. This should be based on the level of completion of the building and other site needs. This helps to determine the amount of space required. It will also identify special material protection requirements and total storage time. For items with long lead times, it may be possible to postpone delivery until shortly before they are needed on site. This helps minimize the risk of damage. Access to materials within the storage/staging area should also be considered. Heavy items requiring equipment to handle them should be placed where they can be accessed without damaging other items.

After the materials inventory is complete, each item in the inventory should be assessed in terms of potential threats that should be guarded against. *Table 9* lists potential threats to be consid-

Table 8 Materials Management Needs Inventory

Material Type	Quantity (Unit)	Task/Time Period	Source	Lead Time	Storage Needs	Comments
Foundations/Substructure						
Concrete						
Steel						
Other...						
Superstructure						
Concrete						
Steel						
Wood						
Other...						
Building Enclosure/Envelope						
Sheathing						
Insulation						
Exterior finish						
Doors/windows						
Other...						
Interior Construction and Finishes						
Drywall						
Wood						
Doors/windows						
Paint						
Carpet						
Tile/rigid flooring						
Ceilings						
Other...						
Services						
Plumbing						
Electrical						
HVAC						
Fire suppression						
Other...						

ered for each material in the material inventory, along with mitigation measures.

Projects pursuing LEED certification, particularly EQ Credit 3: Indoor Air Quality Management, should pay special attention to potential contamination of products. Provisions should also be made in the schedule to ensure that off-gassing materials have adequate time to release chemicals before inspection begins.

Excess materials at the conclusion of each task should be considered in the materials management plan and also addressed specifically in the project waste management plan.

4.3.5 Materials Management Plan Implementation

Implementing the materials management plan requires attention throughout the project. Key considerations include the following:

- *Coordinating product delivery to manage congestion* – Delivery times should be arranged when possible to avoid conflicts with other site traffic. Products that require equipment for offloading from delivery vehicles must also have adequate space to reduce the potential for damage to other products in storage.
- *Minimizing the need to move products on site* – The risk of physical damage is increased every time a product is handled. Ideally, products would be installed right off the truck through just-in-time delivery. For most products, however, this is impractical. Advance planning can help to ensure that products are not stored in areas where they will interfere with concurrent tasks.
- *Minimizing risk of schedule disruption* – Managing this risk is critical for products with long lead times or from unfamiliar sources. Additional time may need to be built into the procurement schedule to ensure that construction is not delayed. Arrangements should be discussed with suppliers to delay taking control of products until close to their installation time.
- *Storage of excess materials to preserve value* – Careful storage of excess materials at the end of installation is important to preserve their value. Products to be taken back by vendors or suppliers are especially important. If donation to charity or freecycling will be used, it makes sense to accumulate excess materials and dispose of them together rather than individually. This requires setting aside a space for excess materials where their value can be preserved.

Table 9 Threats and Mitigation Measures for Material Storage

Threats to Material Integrity	Mitigation Measures
Moisture Exposure to precipitation Excess humidity Absorption from ground contact	• Moistureproof indoor product storage • Placement to allow ventilation • Preventing ground contact • Adequate covering • Active ventilation/heating
Photodegradation Exposure to UV radiation	• Indoor product storage • Adequate covering • Organized laydown yard
Material Security	• Indoor product storage • Protected/locked storage
Temperature Fluctuation	• Indoor product storage • Active ventilation/heating
Physical Damage By equipment during handling By equipment while stored Improper orientation/support	• Indoor product storage • Adequate support • Following manufacturer stacking/protection recommendations
Contamination Exposure to spills Exposure to dust Absorption of contaminants from surrounding materials	• Separate storage of absorptive items from potential contaminants • Adequate covering • Sealed openings • Active ventilation • Cleaning before installation

4.4.0 Waste Management

Managing solid waste on the construction site creates tangible, visible benefits. Psychologically, dumpster contents form a first impression of how green a project is. Although there are more impactful ways to increase project sustainability, effective waste management is a good place to start.

4.4.1 Waste-Generating Construction Processes

Demolition debris and construction scrap material are the most obvious sources of waste on the construction site. However, significant quantities of waste come from other sources that could be avoided, including the following:

- *On-site fabrication of building components* – Fabrication on site increases overall waste rates. The likelihood of reusing scrap materials is much less on a job site than in a centralized facility where similar products are regularly made.
- *Inefficient procurement and transport of products to site* – Purchasing materials in bulk can often avoid significant packaging waste. However, it may also result in excess product on site that is easier to dispose of than return to the manufacturer or reuse.
- *Careless materials management* – Poor storage and staging practices result in materials that are wasted due to physical damage, exposure to moisture or weather, or other damage.
- *Poor coordination of subcontractors* – Failure to coordinate the work of subcontractors can result in damage to installed systems or a need to rework finishes.
- *Poor site layout* – Inefficient location of scrap storage, cutting areas, or disposal areas makes it easier to pull a new piece of material from the stack instead of searching for a scrap piece that will work.
- *Poor protection of installed work* – A lack of effort to protect installed systems, especially finishes, can require that damaged systems be torn out and reinstalled. This not only generates waste, but also increases project cost and extends the completion schedule.

Each of these practices increases the quantity of solid waste generated on a project. Some occur due to a desire to save time or labor costs. Others can actually increase installation time, such as poor protection of installed work. Managing these trade-offs is an important function of the Sustainable Construction Supervisor.

4.4.2 Options for Waste Reduction and Diversion

Many opportunities exist for reducing the amount of solid waste from a project. Additional opportunities exist for diverting unavoidable waste to better destinations. When preparing for construction and procuring products and materials, significant waste can be avoided in the first place by doing the following:

- *Prefabricating components in the shop instead of on site* – This can optimize raw material use and encourage greater use of scrap material, since it is located where similar products will be made in the future.
- *Designing for modular construction* – This reduces waste by sizing building elements to fit standard material sizes, thus reducing the need to cut and customize parts. This approach can also save significant construction time.
- *Ordering materials cut to size* – As with shop fabrication, this encourages centralized cutting, which increases reuse of scrap for future projects and reduces mistakes.
- *Working with manufacturers to reduce packaging waste* – This includes manufacturer takeback of reusable packaging as well as the use of compostable or biodegradable materials.

During construction, process waste can be reduced using the following methods:

- *Managing procurement from vendors* – Requesting that products be shipped with returnable packaging eliminates the need to dispose of the packaging as waste. Ask suppliers to take back or buy substandard, rejected, or unused items.
- *Careful excavation planning* – Balancing project cut/fill requirements reduces the need to dispose of soil spoils off site (*Figure 77*). Stockpiling topsoil for reuse also reduces the need to import new soil at the end of the project.
- *Limiting site disturbance* – This reduces the amount of land-clearing debris generated during the project and also minimizes the cost of re-landscaping at the end of the project.
- *Effective materials management* – This ensures that materials are not damaged while on site *before* they are installed. A staging plan that reduces material movement also reduces the chance for damage.
- *Careful coordination of subcontractors* – Properly sequencing installation of products, especially finishes, can reduce the chance of material damage after installation.

201F77.EPS

Figure 77 Careful excavation planning can minimize the need to transport soil off site.

- *Protection of installed work* – Taking the time to protect finished work or install coverings over installed work reduces the likelihood of damage and need for rework.

In projects requiring removal of existing buildings or systems, waste can be reduced using the following methods:

- *Engaging salvage organizations* – This provides an opportunity for companies with established markets to recover useful products from the building prior to demolition activities.
- **Deconstruction** *instead of demolition* – This involves carefully taking apart building systems so that they can be reused (*Figure 78*).
- *Protection of remaining installed work* – Sealing off areas where demolition is occurring reduces the likelihood of damage to materials and systems that are not part of the project.

For unavoidable waste generated during demolition, renovation, or construction, waste can be diverted from landfills or incinerators by using the following methods:

- *Capturing recyclable materials* – This involves diverting recyclable materials from the solid waste stream either through on-site separation or comingling (combining) for off-site separation (*Figure 79*). Recycling opportunities differ by location based on the availability of facilities, market value, and other factors.
- *Composting* – This involves segregating biodegradable wastes, such as certain types of packaging or land clearing debris, for diversion to off-site composting facilities (*Figure 80*). The effectiveness of this option depends on the types of waste generated and the availability of local facilities.

201F78.EPS

Figure 78 Materials like these plumbing fixtures can be recovered for salvage and donated to charities like Habitat for Humanity.

201F79.EPS

Figure 79 Steel is an easily recycled construction material.

201F80.EPS

Figure 80 Municipal composting facilities such as this can accept organic wastes.

What's wrong with this picture?

201SA14.EPS

On-site disposal options may also be available for certain types of waste, including the following materials:

- *Biodegradable materials* – Certain types of packaging, limited amounts of wood waste, and site-clearing and landscaping debris can be processed on site for use as landscaping mulch.
- *Drywall* – In limited quantities, drywall can also be applied on site as a soil amendment after being processed through a grinder or chipper. Whether this is a good idea depends on soil characteristics.
- *Waste concrete* – Can be used as subbase for paved areas or foundations, or also as riprap for stormwater management or soil retention. It can also be processed on site using crushing equipment for reuse as aggregate.

4.4.3 Developing a Waste Management Plan

A waste management plan takes time to create. It is a good investment for a construction company. As landfill space decreases, waste disposal costs will continue to rise. Increases in the cost of raw materials will also make recycling more economically attractive. Waste recycling is already economically profitable in some parts of the U.S.

Three fundamental activities are part of waste management planning:

1. *Waste stream assessment* – The first step involves reviewing the project to estimate the likely types and quantities of waste associated with each project task. This includes items to be salvaged, recycled, or composted. Identifying the period in the schedule when these wastes will occur is also necessary. Knowing the timing of each waste stream will help to plan site layout. It will also help to ensure that no more dumpsters are on site than are needed at a particular point. For instance, projects that recycle drywall waste do not need a drywall dumpster on site until interior construction begins. Likewise, a concrete waste dumpster is not likely to be needed after demolition is finished. The Sustainable Construction Supervisor will be responsible for coordinating waste management activities on site. *Table 10* provides a template for organizing this information.
2. *Local market evaluation* – The second step is to investigate the local and regional market for recycling and salvage. This will determine potential destinations for each waste stream. It will also establish a value for each type of waste. This evaluation should include waste haulers, recyclers, and on-site waste subcontractors. Other market options for handling waste streams, such as Habitat for Humanity or nonprofit organizations, should also be considered.
3. *Cost/benefit analysis of options* – The third step is to evaluate the costs and benefits of options for each waste stream. The nuances of the project will also determine what is feasible in terms of space constraints for dumpsters.

The cost/benefit analysis will suggest the best approach for the individual project conditions. Direct costs, such as landfill and hauling fees, should be taken into account. Also indirect costs, such as increased management time and labor costs, must also be considered. The following considerations are also important:

- Likely diversion rates
- Project waste diversion goals
- Potential liabilities
- Schedule constraints

Four major categories of options include the following:

1. *On-site segregation* – This option includes all scenarios where waste is sorted into multiple dumpsters located on site. Subcontractors are commonly required to sort their own waste as part of their scope of work. In some cases, a housekeeping subcontractor takes care of sorting and housekeeping on site. Bins or dumpsters may be kept in a centralized location, or may involve multiple smaller bins located near work areas. The latter option is likely to result in a higher diversion rate, but also requires additional labor to consolidate and empty the bins.

Table 10 Waste Stream Assessment Template

Project:				Diversion Method				
Material Type	Expected Quantity (unit)	Generated by (Task)	Time Period	Recycle – To Facility	Recycle – To Salvage	Recycle – On-Site	Dispose – To Facility	Comments
Mixed Materials								
Mixed C&D waste								
Separated Materials								
Asphalt								
Brick/masonry								
Concrete								
Lumber								
Composite wood								
Drywall								
Roofing materials								
Cardboard								
Paper								
Polystyrene foam								
Plastic								
Metals								
Gravel								
Soil								
Compostables								
Other								

2. *Comingling for off-site separation* – This option is available in areas where centralized construction and demolition waste sorting facilities exist. All on-site wastes are deposited by subcontractors into a common bin or dumpster. This dumpster is periodically removed and sorted in an off-site facility for recycling. A specialty waste management subcontractor generally provides this arrangement. It often results in higher diversion rates. The subcontractor provides required documentation for LEED credits. The cost of this option depends on the types and quantities of wastes to be generated. It is also a function of the market conditions. Waste management companies offering this service have well-established market connections. They are able to consolidate waste streams to achieve higher value of recycled materials.
3. *On-site processing* – This option may be appropriate for some waste streams, such as land clearing debris or large-scale concrete waste. It involves contracting specialized equipment to come on site to deal with a single type of waste. This waste will then be re-used on site, such as crushed concrete for aggregate, or wood waste for mulch.
4. *Salvage/deconstruction* – If the project involves work with existing facilities on site, salvage or deconstruction may be economically desirable to recover useful materials. Specialized firms can be employed to salvage reusable products. Often this is in exchange for the value of those products. Demolition subcontractors often provide this service prior to demolishing the building. Capturing documentation of the fate of these materials is important for LEED certification.

Often, projects will involve a hybrid of these approaches based on local market conditions and the specific nature of waste streams involved. The following is a list of key questions to ask salvage or recycling organizations when investigating the local market for recycling and salvage:

- What materials do you accept?
- Are there minimum or maximum quantities that are acceptable?
- Will you pick up the materials at the job site?
- Is there a charge for pick-up service or hauling?
- Who do I call to arrange for pick-up?
- How much lead time is required?
- If I must make separate hauling arrangements, where and how should materials be delivered?
- Will you provide containers?
- Do you charge a container rental fee?
- Must containers be protected from moisture?
- Will you pay for the materials?
- What potential contaminants are problematic?
- What conditions will cause a load to be rejected?
- Can you provide proof of waste diversion by weight or volume?
- What signage or information can you provide to ensure proper separation?

The first step is inventorying the likely waste streams from the project. Next, check the local market. The last step in developing a waste management plan is to evaluate the costs and benefits of each option. Spreadsheet tools available online (for example, see http://your.kingcounty.gov) can help evaluate the costs of disposing versus recycling different materials for your project. Analysis may also include non-monetary costs and benefits. This may include things such as the likelihood of meeting LEED diversion goals and the corporate environmental image.

4.4.4 Waste Management Plan Implementation

Successful waste management goes beyond just creating a good plan. Actively monitoring and controlling implementation is necessary for success in meeting waste diversion goals. Key considerations include the following:

- *Contamination of dumpsters* – Incorrect waste sorting on site can lead to contaminated loads. These loads may be rejected or discarded by recyclers. Depending on the market value of the material, the recycler may reject the load and send it back to the project. The recycler may also discard it in a landfill. In either case, the goals of waste diversion are not being met. Clear responsibility should be established for each type of waste sorted on site. The Sustainable Construction Supervisor should take measures to ensure that segregated dumpsters are not contaminated. Subcontractors should receive training on proper sorting methods. They should be clear on what is and what is not acceptable for each dumpster or bin. Signs should be clearly posted in each disposal area to communicate proper behaviors. The Sustainable Construction Supervisor should also be available to answer the questions of workers on site. Keep in mind that it is far better to discard a small amount of potentially recyclable waste than to risk rejection of a whole load due to contamination.
- *Timing of dumpster pulls* – Timing of dumpster removal and replacement is worth considering. The timing should be considered from the perspective of site congestion, equipment idling, and transportability. Requirements for on-site sorted waste dumpsters should be established by the recyclers and haulers. These requirements should be carefully observed by workers on site. Some haulers require notification to schedule dumpster pulls. Other haulers will pull on a regular schedule. In general, hauling full instead of partial dumpsters is more efficient from a transportation standpoint. However, lead time may be required to have the dumpster replaced. During this time some recyclable material may go into the general dumpster because the recycling dumpster is full. The Sustainable Construction Supervisor should obtain a clear understanding of each recycler's or hauler's policies and practices to prevent this problem.

What's wrong with this picture?

201SA15.EPS

- *Penalties for illegal dumping and unsecured loads* – Many municipalities have fines and penalties for illegal dumping of waste and/or transporting unsecured loads. Such fines can be on the order of $5,000 in some areas if an item falls off the vehicle and causes bodily injury. The responsible party may also face criminal charges. Ensure that hauling and disposal companies are reputable, insured, and bonded.
- *Documentation availability and tracking* – If LEED certification is an aim of the project, proper documentation of waste destinations and quantities is critical for calculation purposes. Some waste management subcontractors will manage the entire process and guarantee a particular level of waste diversion for LEED purposes. Other situations may require the Sustainable Construction Supervisor to play a more critical role. Clear roles and responsibilities for documentation and tracking should be established as part of the waste management plan. Even if waste management is subcontracted, the Sustainable Construction Supervisor should periodically check to be sure that LEED requirements are being met.
- *Liability of salvage entities* – If salvage or deconstruction is to be employed as part of the project's waste management strategy, certain items need to be covered in the contract. The liability associated with the operation should be clearly apportioned in the contract documents before salvage begins, especially in cases where non-profit organizations or laypersons are involved in salvage or deconstruction. In these cases, waivers of liability for the general contractor may be desirable. Partnering with non-profit organizations like Habitat for Humanity and local tree transplant/salvage operations is well worth considering as a waste management strategy. Not only does the project reduce waste, but also social goals are met and public relations improved. There may even be tax deductions available for donations of products to these groups. However, liability issues may damage an otherwise positive effort if not properly taken into account.

4.5.0 Indoor Air Quality Management

Actions during construction may have a significant influence on air quality in the finished building. Many opportunities exist to prevent potential problems through careful construction practices. An **IAQ management plan** is essential to ensure that the resulting building does not suffer from easily avoidable problems later.

Did You Know?

Depending on the market value of the material and its physical and chemical properties, different levels of contamination may be acceptable. For example, contamination of metals is generally less of a problem than for other materials, since recycling of metals involves extremely high temperatures that incinerate many contaminants. Plastics, on the other hand, are much more susceptible to problems from mixing and contamination. Even plastics of the same number may not be chemically compatible. For instance, plastics are more desirable for recycling if they have been blow-molded, such as milk jugs and soft drink bottles. Injection-molded plastics, such as plastic clamshell containers, melt at a lower temperature and are not compatible with blow-molded plastics. The presence of a number and the recycling symbol on a piece of plastic does not necessarily mean it can be recycled in practice.

4.5.1 Construction Effects on IAQ

Construction can be responsible for introducing contaminants into the indoor environment. Construction activities, such as material cutting and spraying finishes, introduce potential contaminants to the building. So do pouring concrete, painting, and installing adhesives. Simply moving materials into and out of interior spaces can introduce contamination. This occurs through offgassing of those materials.

What's wrong with this picture?

201SA16.EPS

Many opportunities exist to reduce these problems. Reducing contaminants in the indoor environment has a number of benefits. These include improved comfort and performance and reduced health risks. These benefits are for both workers and future building occupants. Health risks can be reduced by choosing less harmful materials and by changing the way they are installed. Many of the measures discussed earlier for pollution prevention can improve indoor air quality as well. Wet processes instead of dry processes help to suppress dust and particulates. Workers should also wear personal protective equipment and employ dust collection when dust is generated through construction processes such as sanding, finishing, and welding.

Careful attention to pollution prevention can also have economic benefits. Capturing or avoiding contaminants in the first place can reduce the time and cost required to clean them up later. Cleanup often is required at the end of construction when the schedule is already tight. Reducing cleanup requirements can make the difference between a project that ends on time and one that is late.

4.5.2 Options for IAQ Management

Multiple options exist for improving indoor air quality. The Sheet Metal and Air Conditioning Contractors National Association (SMACNA) has developed a standard for managing indoor air quality. These standards are applicable during construction, demolition, or renovation of occupied spaces. Chapter 3 of this standard focuses on control measures and guidelines to be used during construction. These include the following:

- *HVAC protection* – Protect all HVAC equipment from dust and odors during construction. This should happen for both temporary and permanent equipment. Before installation, all duct and equipment openings should be sealed with plastic. HVAC equipment operated during construction should be protected with filters at all return air openings or on the negative pressure side of the system. Unducted plenums in the construction work zone should leave ceiling tiles in place and seal off return air grilles. The contractor should check for leaks in return ducts and air handlers. These should be sealed promptly to prevent introduction of contaminants. Mechanical rooms should not be used for storage during construction. All filtration media should be replaced whenever dirty or clogged during construction and again before the building is occupied.
- *Source control* – Use low-emitting materials wherever possible. Where materials containing VOCs must be used, proper control measures should be employed. For instance, areas where such materials are stored or installed should be isolated from other areas. These spaces should be properly ventilated. Equipment that generates exhaust fumes should also be isolated or avoided if possible. This includes exhaust from idling vehicles near air intakes for the building. Dust collection systems should be used on all equipment used for cutting or sanding.
- *Pathway interruption* – Construction activities should be physically isolated from clean or occupied areas. This can be accomplished with temporary barriers. This can be as simple as using plastic sheeting and tape. Isolated areas should be kept under negative pressure to keep contaminants within the space. Ventilation with outside air should be used. This is especially important during construction tasks that generate air pollution. All absorptive materials should be protected both before and after installation to prevent contamination.
- *Housekeeping* – Proper maintenance and cleaning should be regularly undertaken during construction. This is especially important after spaces have been enclosed. This provides a safer work environment. It also prevents problems for future occupants. Porous materials should be protected or cleaned with dry methods instead of wet methods. This prevents exposure to moisture. High-efficiency vacuum cleaners can be used to clean dusty areas without spreading dust further.
- *Scheduling* – Sequencing of construction activities can be used to prevent dust and debris from contaminating previously installed materials. If buildings are occupied during regular business hours, dirty or odorous tasks should be scheduled during off hours to reduce impacts to occupants. Adequate time should be included in the schedule for building flush-out and/or air quality testing after construction is complete. The schedule should also include a reminder to replace all filtration media prior to occupancy.

4.5.3 Developing an IAQ Management Plan

The process of developing an IAQ management plan is similar to planning for materials management and pollution prevention. In fact, these plans have a number of elements in common. They all should be reviewed and coordinated before the project begins. When

What's wrong with this picture?

201SA17.EPS

developing an IAQ management plan, the first step is to identify potential threats to IAQ. These are typically associated with specific construction tasks. A template similar to *Table 8* can be used to assess the risk. These risks are associated with installation of specific products, materials, and systems on the project. Specific construction processes should also be reviewed to identify potential threats. Some of these process threats have already been identified and planned for in the Pollution Prevention Plan. IAQ management generally focuses in more detail on construction processes that happen after the building has been enclosed.

The second step for an IAQ management plan is to identify options to reduce or eliminate threats. In general, options that prevent pollution are preferable to options that try to fix it after the fact. Specific measures recommended by SMACNA should be reviewed as potential options and incorporated into the plan.

The third part of IAQ management planning is to evaluate solutions in terms of their costs and benefits for the project. Both the hard costs of each option and nonmonetary costs and benefits should be included. Reduced callbacks, owner satisfaction, and contractor reputation are all worth considering. Reduced environmental liability is also important. Contractors can be involved in litigation for sick building syndrome long after the project is complete. Careful attention to IAQ and documentation of efforts can be invaluable in reducing liability if problems occur later in the building's life cycle.

4.5.4 IAQ Management Plan Implementation

All subcontractors on the project should be made aware of IAQ goals for the project. They should be given copies of the IAQ management plan. Periodic inspection by the Sustainable Construction Supervisor can help subcontractors comply. When determining how to implement provisions of the IAQ plan, key considerations include the following:

- *Implications of IAQ testing versus flush-out* – If pursuing LEED certification, the team may elect to conduct indoor air quality testing instead of taking time in the schedule for flush-out. If this is the case, the Sustainable Construction Supervisor must be especially vigilant to schedule activities and material installation immediately before testing. Since testing occurs immediately prior to occupancy, this is a period when punchlist activities are often occurring. In the zeal of trying to complete the project and address punchlist issues, subcontractors sometimes take shortcuts. Any activity that would disturb or contaminate existing finishes, or any last-minute material installations, should be very carefully reviewed using the guidelines in this module to ensure they will not threaten the outcome of the IAQ tests.
- *Ventilation with temporary versus permanent equipment* – The manufacturer's warranty period on permanent equipment generally begins as soon as the equipment is installed and started up. If permanent equipment is used during construction, the manufacturer's warranty may run out before the contractor's warranty after the building is occupied. This may expose the contractor to potential liability. Using temporary equipment may be preferable to avoid this possibility.

What's wrong with this picture?

201SA18.EPS

- *Coordination with other project implementation plans* – The construction IAQ management plan likely has common elements with other implementation plans. These include the sustainable operations and pollution prevention plan, the materials management plan, the commissioning plan, and the waste management plan. These plans should be developed to be complementary and coordinated. This helps to prevent confusion and ensures that critical responsibilities are managed correctly.

4.6.0 Commissioning

The aim of commissioning is to ensure that building systems perform according to the design intent and owner's requirements (*Figure 81*). Contractors are involved in delivering and installing building systems to create a complete facility. They represent a critical link in the commissioning chain.

4.6.1 The Commissioning Process and Intent

Depending on the project delivery method that is being used, the Sustainable Construction Supervisor may be involved in commissioning activities throughout the project delivery process. For instance, in design-build projects, the construction team may be invited to participate as part of pre-construction activities. This may include project planning or design. In these cases, the Sustainable Construction Supervisor can contribute knowledge and expertise on how construction means and methods will affect system delivery and performance. These may contribute to the owner's project goals and requirements. The commissioning authority (CxA) may also be a member of the contractor's organization, particularly if the project is being delivered by a construction management at risk delivery method.

201F81.EPS

Figure 81 Commissioning involves review of design intent versus performance.

The overall commissioning process parallels the project delivery process, as shown in *Figure 82*.

4.6.2 Construction-Related Commissioning Activities

The primary involvement of the Sustainable Construction Supervisor with commissioning activities will occur during planning and implementing the construction process. Specific roles played by the Sustainable Construction Supervisor during this time may include the following:

- Reviewing and understanding commissioning requirements in the construction documents and commissioning plan
- Preparing a commissioning milestone schedule
- Undertaking any responsibilities specifically assigned to the construction supervisor in the commissioning plan
- Informing and coordinating responses of subcontractors to commissioning requirements in the construction documents or commissioning plan

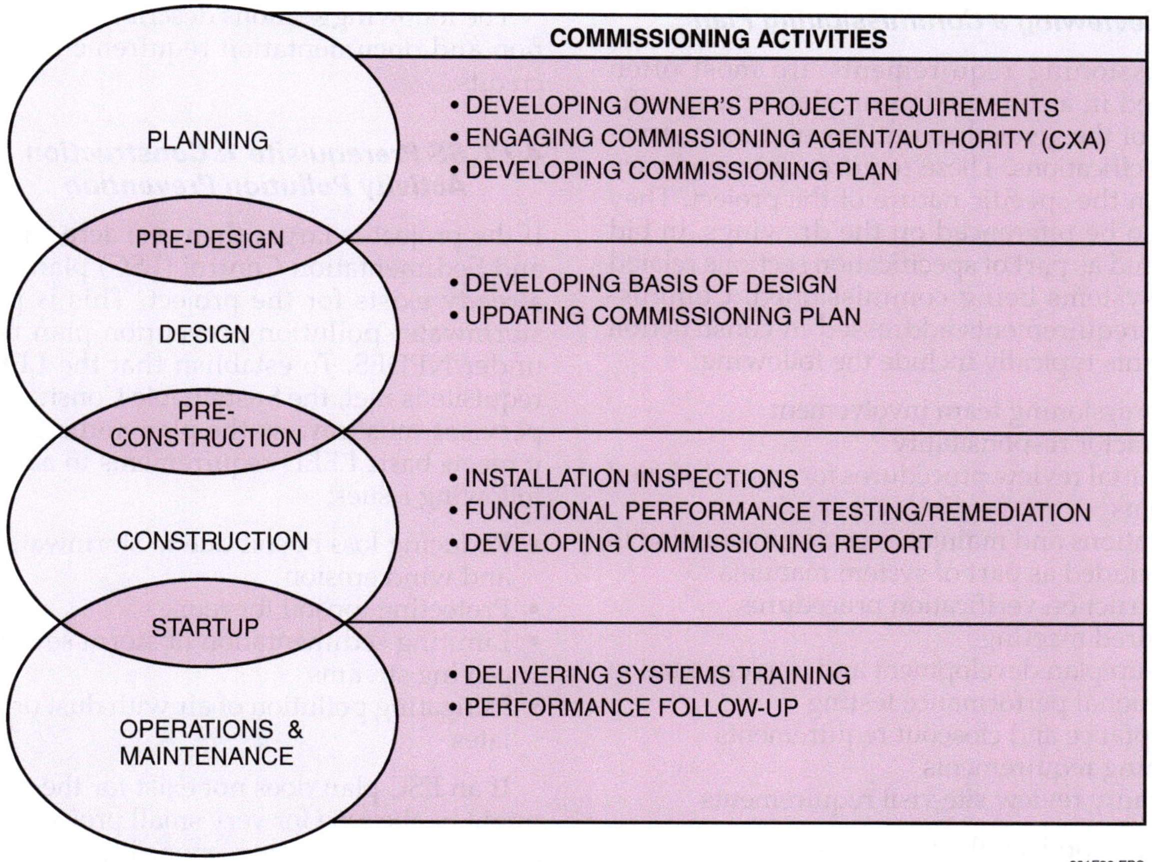

Figure 82 Typical commissioning activities by project phase.

- Coordinating the presence of the commissioning authority with scheduled construction activities such as system installation and performance testing
- Notifying the commissioning authority of any relevant changes in the project scope or schedule
- Verifying that the commissioning authority has met the requirements of the commissioning plan
- Providing copies of submittals and documentation pertaining to commissioned systems to the commissioning authority for review
- Coordinating training on commissioned systems as part of project closeout

Working with the Sustainable Construction Supervisor, the commissioning agent will be involved with installation inspection. They will also be involved with performance testing. A purpose of these activities is to ensure that equipment is properly installed. Another purpose is to ensure that the equipment is able to perform as desired.

Installation inspection of commissioned systems often occurs at start-up of individual system components. The commissioning agent will follow prefunctional checklists or start-up and checkout forms. These forms serve to document the start-up process. Depending on the requirements of the commissioning plan, the installing contractor might complete the forms instead of the commissioning agent. The purpose of these inspections is to discover any improperly installed components prior to performance testing.

System performance testing, also called functional performance testing, occurs after all system components have been installed. All components within each system to be commissioned should be fully ready for operation. These systems should be able to operate under part-load or full-load conditions. Performance testing then involves testing each process in the sequence of operations. This sequence includes start-up, shutdown, capacity modulation, emergency and failure modes, alarms, and interlocks to other equipment.

The purpose of the test is to simulate conditions under all modes of operation. This is done to ensure the equipment performs as expected. The outcomes of the test will identify performance problems that need to be fixed before the system will be acceptable to the owner. The Sustainable Construction Supervisor will work with other members of the project team to develop and implement a plan to fix any identified problems.

4.6.3 Reviewing a Commissioning Plan

Commissioning requirements are most often described in a commissioning plan or a specific section of the general conditions of the construction specifications. These requirements may vary based on the specific nature of the project. They may also be referenced on the drawings, in bid forms, and as part of specification sections related to the systems being commissioned. Commissioning requirements addressed in construction documents typically include the following:

- Commissioning team involvement
- Contractor responsibility
- Submittal review procedures for commissioned systems
- Operations and maintenance documentation to be included as part of system manuals
- Construction verification procedures
- Required meetings
- Start-up plan development and implementation
- Functional performance testing
- Acceptance and closeout requirements
- Training requirements
- Warranty review site visit requirements

For each project, the Sustainable Construction Supervisor should carefully review the project commissioning plan and all construction documents to identify specific contractor requirements. A key contractor submittal often required on green projects is the commissioning milestone schedule. This schedule demonstrates to the project team that all construction-phase commissioning requirements have been identified and planned. It also provides a basis for coordinating installation inspections and a basis for the functional tests during installation of commissioned systems.

4.7.0 Calculating and Documenting Process-Related LEED Credits

Depending on the nature of the project, the Sustainable Construction Supervisor may play various roles in the LEED certification process. Construction-phase credits depend on proper documentation to meet certification requirements. These credits include, but are not limited to the following:

- SS Prerequisite 1: Construction Activity Pollution Prevention
- SS Credit 5: Site Development
- EA Prerequisite 1/Credit 3: Fundamental and Enhanced Commissioning
- MR Credit 2: Construction Waste Management
- EQ Credit 3: Indoor Air Quality (IAQ) Management Plan

The following sections describe LEED certification and documentation requirements for these credits.

4.7.1 SS Prerequisite 1: Construction Activity Pollution Prevention

If the project is larger than one acre, an Erosion and Sedimentation Control (ESC) plan probably already exists for the project. This is part of a stormwater pollution prevention plan required under NPDES. To establish that the LEED prerequisite is met, the Sustainable Construction Supervisor must review the plan and confirm that it meets basic LEED requirements to address the following issues:

- Reducing loss of soil due to stormwater runoff and wind erosion
- Protecting topsoil for reuse
- Limiting sedimentation of storm sewers or receiving streams
- Mitigating pollution of air with dust or particulates

If an ESC plan does not exist for the project, as might be the case for very small projects, the general contractor may be responsible for developing one. Details on how to construct an ESC plan are outside the scope of this module. However, more information is available online through the U.S. EPA's Office of Water. Specific requirements and guidelines are available at http://cfpub.epa.gov.

LEED documentation for the prerequisite includes an erosion and sedimentation control drawing. It also requires a written ESC plan (see checklist). These should include specifications that detail best management practices to be used on the project. They should also specify the responsible party for each aspect of the plan.

The Sustainable Construction Supervisor must also verify and confirm that the plan has been properly implemented. This involves maintaining erosion and sedimentation control measures on site throughout construction. They should also photograph each of these measures (*Figure 83*). They should also prepare a copy of the site plan showing the location of each photograph as part of LEED project documentation. Photographs should be taken when these measures are initially installed. They should also be taken periodically throughout the project to demonstrate that they are maintained and functioning properly. Photographs should be dated and time stamped. This will allow them to be compared to the project log and correlated with weather events.

The Sustainable Construction Supervisor may also keep an inspection log to document the

201F83.EPS

Figure 83 Protecting stormwater inlets is required for stormwater management.

ongoing condition of control measures. This log should include the following:

- Dates of inspection
- References to any photographs taken
- Control measure(s) inspected
- Condition of control measures and notes on causes of problems
- Corrective actions taken in response to problems

At the end of the project, the general contractor is responsible for removing any temporary erosion and sedimentation control measures used on the site. The contractor is also responsible for cleaning up any residual sediment in the area.

4.7.2 SS Credit 5: Site Development

Sustainable Sites Credit 5 encourages the protection and enhancement of native habitat and open space on building sites. Of the two parts of the credit, *Credit 5.1: Site Development – Protect or Restore Habitat* is of greatest relevance for Sustainable Construction Supervisors. The intent of the credit is to conserve existing natural areas. It is also focused upon restoration of damaged areas to provide habitat and promote biodiversity. For greenfield sites, Credit 5.1 requires that disturbance beyond minimum boundaries be limited during construction (*Figure 84*). Limiting disturbance on previously disturbed sites is not required. However, it is also likely to preserve existing vegetation and soil structure. As such, careful control of site disturbance is recommended in either case.

CHECKLIST
SS Prerequisite 1: Construction Activity Pollution Prevention

The Erosion and Sedimentation Control (ESC) plan must include provisions for:

- ❑ Avoiding soil loss
  - Stormwater runoff
  - Wind erosion
- ❑ Stockpiling and protecting topsoil for reuse
- ❑ Preventing sedimentation from stormwater
  - Storm sewers
  - Receiving streams
- ❑ Preventing air pollution from dust/particulates

LEED documentation includes the following:

- ❑ Erosion and Sedimentation Control (ESC) plan
  - Site drawings
  - Written plan
  - Specifications for Best Management Practices
  - Responsible parties for all plan aspects
- ❑ Verification of proper implementation
  - Date-stamped photographs of control measures
  - Site plan indicating location of photos
  - Inspection log or report
    - Dates of inspection
    - References to photographs taken
    - Control measure(s) inspected
    - Condition of control measures
    - Notes on causes of problems
    - Corrective actions taken in response to problems

What's wrong with this picture?

201SA19.EPS

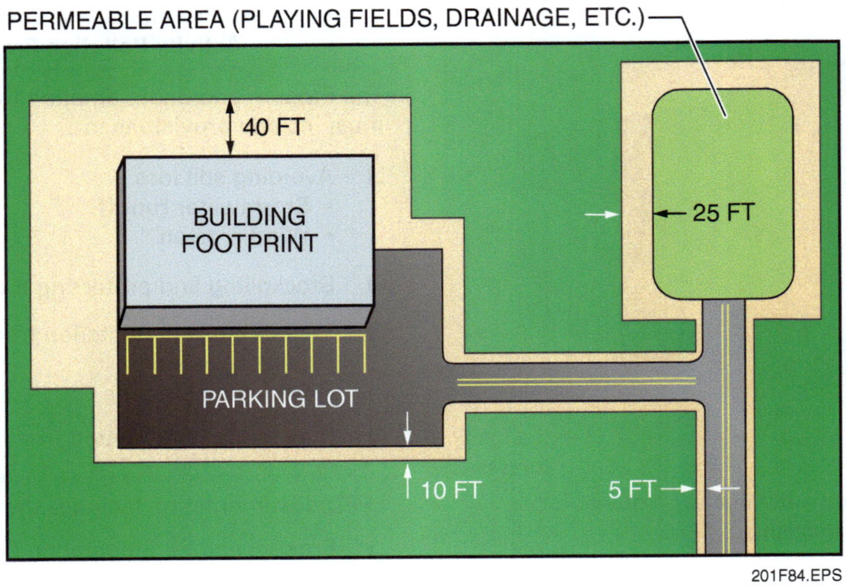

Figure 84 Site disturbance must stay within these limits to meet LEED requirements.

Required documentation for this credit includes a site plan with limits of disturbance clearly indicated. The Sustainable Construction Supervisor is responsible for laying out this boundary on the actual project site and enforcing it. Photos may also be taken to document the proper location of the boundary and limitations of disturbance during construction. These should be indexed and indicated on the site plan to show their location.

The most important aspect of this credit for the Sustainable Construction Supervisor is to meet not only the letter of the requirements but also the intent. Site disturbance should not just be limited to the distances specified by LEED. Instead, it should also carefully take into account natural site features, such as mature trees, that should be preserved. Fencing should be established with sufficient setbacks from natural site features to ensure that desirable areas are actually preserved.

CHECKLIST
SS Credit 5.1: Site Development – Protect or Restore Habitat

Responsibilities of the Sustainable Construction Supervisor include the following:

❑ Establishing limits of disturbance for the project as indicated in the site plan or to meet LEED requirements, such as no more than:
- 40 feet beyond the building perimeter
- 10 feet beyond surface walkways, patios, surface parking, and utilities less than 12 inches in diameter
- 15 feet beyond primary roadway curbs and main utility branch trenches
- 25 feet beyond constructed areas with permeable surfaces that require additional staging areas to limit compaction in the constructed area
  - Pervious paving areas
  - Stormwater detention facilities
  - Playing fields

❑ Designing site operations, laydown and storage areas, and construction traffic corridors to remain within disturbance boundaries

❑ Enforcing disturbance boundaries for subcontractor operations

❑ Supervising habitat restoration activities for disturbed areas

4.7.3 EA Prerequisite 1/Credit 3: Fundamental and Enhanced Commissioning

While most of the LEED documentation associated with commissioning is the responsibility of the commissioning agent, the contractor (represented by the Sustainable Construction Supervisor) plays a key role. Specifically, the contractor is required to prepare a commissioning milestone schedule. This schedule coordinates the installation inspections and functional performance testing of key systems as they are being installed. The contractor is also responsible for ensuring that the commissioning agent has access to all construction-phase submittals. These include such items as shop drawings, laboratory tests of relevant products, etc.

In addition to standard review of contractor submittals conducted by the design team, the project commissioning authority may also review submittals. The commissioning authority double-checks submittals. Their role is to identify any issues that might lead to rework or change orders. Specifically, the commissioning agent checks that items in submittals conform to the owner's requirements and basis of design, fulfill the owner's operations and maintenance requirements, and meet minimum performance testing requirements.

The commissioning agent is responsible under EA Credit 3 to prepare an operations manual for each commissioned system. This manual is separate from the operations and maintenance (O&M) manual typically provided by the contractor. This manual focuses on operating the equipment to meet performance requirements. It also covers interactions of each system with other systems that influence performance. The Sustainable Construction Supervisor should be aware of this requirement. They should coordinate, where possible, in preparing and submitting contractor's O&M manuals for commissioned systems.

The commissioning agent is also responsible under EA Credit 3 for providing operator training for commissioned systems. The nature and scope of this training will be described in the commissioning plan. The Sustainable Construction

CHECKLIST

EA Prerequisite 1: Fundamental Commissioning/EA Credit 3: Enhanced Commissioning

Responsibilities of the Sustainable Construction Supervisor include the following:

- ❑ Identifying responsibilities assigned to the construction supervisor and subcontractors in the commissioning plan:
  - Required meetings
  - Submittal review procedures
  - Construction verification procedures
  - Start-up procedures/installation inspection
  - System/functional performance tests
  - Acceptance/closeout requirements
  - Coordination of systems and O&M manuals (EA Credit 3)
  - Training requirements (EA Credit 3)
  - Warranty review requirements (EA Credit 3)
- ❑ Preparing a milestone schedule for commissioning activities
- ❑ Coordinating responses of subcontractors to commissioning requirements
- ❑ Providing copies of all contractor submittals for commissioned systems to the commissioning authority (CxA):
  - Shop drawings
  - Laboratory test results of materials
  - Other
- ❑ Coordinating the presence of the CxA with the construction schedule for commissioned systems:
  - Scheduling installation inspections
  - Scheduling performance tests
  - Scheduling O&M manual delivery
  - Scheduling operator training
- ❑ Notifying CxA of any relevant changes in the project scope or schedule
- ❑ Verifying that CxA has met the requirements of the commissioning plan
- ❑ Providing copies of system submittals and operations and maintenance documentation pertaining to commissioned systems to CxA for review

Supervisor should be aware of this requirement. They should also be sure that it is included in the start-up schedule. Such training often includes instruction on how to use O&M manuals. The Sustainable Construction Supervisor should be prepared to coordinate delivery of these manuals with any planned training by the commissioning agent, subcontractors, and suppliers.

4.7.4 MR Credit 2: Construction Waste Management

The construction waste management credit requires both a plan for construction waste management and a record of what actually happened with the project's waste. Often, both of these requirements are the responsibility of the Sustainable Construction Supervisor.

The construction waste management plan establishes goals for the project's waste diversion. It should also include an inventory of the types and quantities of waste to be diverted on the project. The strategies or measures to be used for diversion, such as on-site separation, should also be detailed. Specific descriptions of the who, what, when, where, and how should be provided for each waste stream to facilitate planning. For instance, what will be done with cardboard waste from the project (*Figure 85*)? When will it primarily be generated? How should it be stored on site to meet recycler requirements? Must it be kept dry and baled? What should be done if it gets wet? Who will be responsible for baling it and arranging for it to be hauled away? Where will it be taken?

201F85.EPS

Figure 85 Corrugated cardboard is a common construction waste that can be easily recycled and has fairly good value for recyclers.

An overall waste management schedule should also be part of the plan if the intent is to separate waste on-site. This schedule coordinates expected waste generation events with waste management responses. Using the schedule, the Sustainable Construction Supervisor can quickly determine likely time frames for scheduling dumpster pulls. This will also help to determine the number and types of dumpsters to be present on site at any point in time. The waste management plan may also include a site plan showing the location of dumpsters and traffic patterns for the site. This is especially important on congested sites or sites with strict limitations on site disturbance.

Recommended documentation for waste diversion rates includes a summary log of construction waste and documentation of what happened to each type of waste. The summary log should include the following:

- An entry for each time material is removed from the site
- The date of the removal event and the party doing the removal
- The estimated or measured weight or volume of that material
- The physical destination of the material, including the type of destination (landfill, recycler, etc.)

The log should include both diverted waste and waste disposed of normally. This enables calculation of the percent diversion rate at the conclusion of the project. Some types of waste, such as materials captured through deconstruction or salvage, will not have actual weight or volume measurements. These must be estimated using the supervisor's best judgment. Other waste streams, such as dumpster pulls, will have documentation of actual weight by the hauler. The Sustainable Construction Supervisor should keep copies of the waste tipping receipts as well as any other records to support the calculations for this credit.

4.7.5 EQ Credit 3: IAQ Management Plan

Achieving EQ Credit 3.1 requires an IAQ management plan plus documentation of how the plan was implemented during the project. The IAQ management plan must outline how required LEED criteria will be met on the project. It should also assign responsibilities to specific people. The responsible parties for IAQ management during construction will be responsible for documenting compliance with the plan as well as how any problems were addressed during the project.

CHECKLIST
MR Credit 2: Construction Waste Management

Responsibilities of the Sustainable Construction Supervisor include the following:

- ❑ Inventorying waste streams to be generated by the project
- ❑ Conducting market research to determine waste diversion options
- ❑ Analyzing the costs and benefits of options
- ❑ Developing a project waste management plan based on a template:
  - Waste diversion goals
  - Inventory of waste types
  - Destination for waste by type
  - Schedule of waste generation
  - Strategies/implementation measures for waste diversion
  - Responsibilities of individual(s)
  - Site plan with dumpster locations and traffic patterns (optional)
- ❑ Keeping a waste management summary log:
  - Material removal events
  - Dates of material removal
  - Estimated/measured weight or volume
  - Destination of material

EQ Credit 3.1 requires the contractor to design and implement an IAQ management plan for the construction and preoccupancy phases of the building. Three key requirements must be part of the plan:

1. Meeting or exceeding the recommended control measures developed by SMACNA. These measures are documented in Chapter 3 of SMACNA's *IAQ Guidelines for Occupied Buildings Under Construction*, 2nd edition (2007).
2. Protecting absorptive materials and products that are stored on site from moisture damage and other environmental threats.
3. Using filtration media with a **minimum efficiency reporting value (MERV)** of 8 in permanent air handlers used during construction, and replacing those media prior to occupancy (*Figure 86*).

Smoking is also prohibited inside the building and within 25 feet of building entrances once the building is enclosed for projects certified under LEED for Schools.

Documentation for EQ Credit 3.1 includes a written IAQ management plan containing the required elements. The recommended structure for the plan includes the following:

- IAQ goals and scope
- Relevant SMACNA guidelines and controls
- Responsible parties
- Guidance for implementation
- Quality assurance/quality control measures

201F86.EPS

Figure 86 Temporary filters remain in place throughout construction.

The *LEED Reference Guide* contains a brief example of such a plan for reference purposes. The Sustainable Construction Supervisor may not be responsible for developing the plan. However, the supervisor is likely to be responsible for its implementation and quality control.

Documentation for EQ Credit 3.1 also includes a detailed photo log. This log should document the measures taken to comply with the IAQ management plan at various points during construction. Usually at least three digital photos per measure during various stages of construction are recommended.

The Sustainable Construction Supervisor may also keep an inspection log or report to document the ongoing condition of control measures. This log should include the following:

- Dates of inspection
- References to any photographs taken
- Control measure(s) inspected
- Condition of control measures and notes on causes of problems
- Corrective actions taken in response to problems

At the end of the project, the general contractor is responsible for removing any temporary IAQ control measures used during the project.

EQ Credit 3.2 requires either a flush-out or baseline IAQ testing after all finishes and furnishings have been installed but before occupancy. The Sustainable Construction Supervisor may be responsible for maintaining a log of conditions during building flush-out if this path is taken. This should be a daily log that includes the date, building occupancy, outdoor air delivery rates, internal temperatures, humidity, and any other special considerations that might affect IAQ.

4.8.0 Evaluating and Documenting Process Substitutions

Change is a regular and expected part of construction projects. If a project is seeking LEED certification, managing change becomes critical. Some changes to the project delivery process can disrupt the project's LEED goals. The Sustainable Construction Supervisor is responsible for reviewing and evaluating changes and taking actions to ensure that they do not cause problems for LEED certification.

4.8.1 Matching Proposed Changes to LEED Credits

The first step in evaluating potential changes in construction processes is to determine how they might influence specific LEED credits. *Table 11* lists different types of changes that might happen on a construction project and the LEED credits that could be affected by those changes.

CHECKLIST
EQ Credit 3.1: Construction IAQ Management Plan – During Construction

IAQ plan must include provisions for the following:

- ❑ Applying SMACNA recommended control measures:
  - HVAC protection
  - Source control
  - Pathway interruption
  - Housekeeping
  - Scheduling
- ❑ Protecting absorptive materials stored on site from moisture damage
- ❑ Using temporary MERV 8 filtration media during construction
- ❑ Prohibiting smoking inside the building (LEED for Schools)

LEED documentation includes the following:

- ❑ IAQ management plan:
  - Goals and scope
  - SMACNA guidelines and controls
  - Responsible parties
  - Guidance for implementation
  - Quality assurance/quality control measures
- ❑ Verification of proper implementation:
  - Date-stamped photographs of control measures
  - Building plans indicating location of photos
  - Inspection log or report
    - Dates of inspection
    - References to photographs taken
    - Control measure(s) inspected
    - Condition of control measures
    - Notes on causes of problems
    - Corrective actions taken in response to problems

CHECKLIST
EQ Credit 3: Construction IAQ Management

Responsibilities of the Sustainable Construction Supervisor include the following:

- ❑ Developing a project IAQ management plan from a template (Credit 3.1):
  - Goals and scope
  - SMACNA guidelines and controls
  - Responsible parties
  - Guidance for implementation
  - Quality assurance/quality control measures
- ❑ Implementing SMACNA guidelines (Credit 3.1):
  - HVAC protection
  - Source control
  - Pathway interruption
  - Housekeeping
  - Scheduling
- ❑ Protecting materials on site from damage (Credit 3.1)
- ❑ Managing filtration media on permanent air handlers (Credit 3.1)
- ❑ Verifying proper implementation of IAQ plan (Credit 3.1):
  - Date-stamped photographs of control measures
  - Building plans indicating location of photos
  - Inspection log or report
- ❑ Schedule building flush-out or IAQ baseline testing (Credit 3.2)
- ❑ Documenting conditions during flush-out using a daily log (Credit 3.2):
  - Building occupancy
  - Outdoor air delivery rates
  - Internal temperatures
  - Humidity
  - Other considerations
- ❑ Removing temporary control measures at the end of the project (Credit 3.1)

Table 11 Potential Process Changes versus Affected LEED Credits

Proposed Change	SS Prerequisite 1	SS Credit 5	EA Prerequisite 1/Credit 3	MR Credit 2	EQ Credit 3
Equipment Changes	X	X		X	X
Earthmoving/excavation	X	X			
Material cutting				X	X
Application of finishes			X	X	X
Schedule Changes	X	X	X	X	X
Material delivery	X	X			X
Material installation		X	X	X	X
Site Layout Changes	X	X		X	

4.8.2 Determining Equivalence

After the potential impacts to LEED credits have been identified, the next step is to determine whether the proposed change will cause problems for the identified LEED credits. If potential problems may occur, the Sustainable Construction Supervisor should reject the proposed change or bring it to the attention of the project sustainability team for verification.

Examples of changes that could cause problems for LEED goals include the following:

- *Using equipment that requires a larger operating area* – This may mean that goals for limiting site disturbance can no longer be met. It may also mean that locations of ESC measures may need to be adjusted.
- *Using equipment that results in more airborne particulates for cutting or finish installation* – This may require using additional IAQ management measures to ensure that air quality is not compromised.
- *Changes in material delivery schedules* – This may require additional material storage and protection measures to prevent damage to materials. Additional space requirements could change construction pollution prevention measures. Additional storage space might also increase site disturbance.
- *Changes in material installation schedules* – This will require coordination with the commissioning authority to allow for prefunctional and performance testing of commissioned equipment. It may also require disrupting the site layout, rescheduling dumpster pulls, and implementing different IAQ control measures.
- *Changes to site layout* – This could be caused by any number of other changes. It may require adjustment to the ESC plan, adjusting site disturbance goals, and relocating receptacles for construction waste.

4.8.3 Approving and Documenting Changes

When the need for a process change is identified, the Sustainable Construction Supervisor should review the proposed change and determine potential implications for LEED credits and project sustainability. If there is a question regarding influence of the change on LEED credits, the Sustainable Construction Supervisor should consult with the project sustainability team before approving the change. Subcontractors should be made aware of the project's LEED goals and be encouraged to consult with the supervisor to discuss proposed changes.

If changes are approved, addenda should be prepared to implementation plans as necessary. The Sustainable Construction Supervisor may also keep a process change log to document the details of the proposed change. This log should document who requested the change, why the change was requested, what the change entails, which LEED credits may be impacted, and what decision was made. These details can be useful in justifying decisions later in the project.

"Learning how to identify and implement Sustainable Construction techniques will enable construction superintendents and foremen to reduce their impact on the world's resources. Through this course, Sustainable Construction Supervisors will learn how to build better buildings."

Dave Conlon, LEED AP, Senior Project Manager, Hardin Construction

Review Questions

1. An example of how construction can impact a neighboring site due to emissions in the air is _____.
   a. chemicals migrating offsite
   b. pollution of groundwater
   c. vibrations from machinery
   d. migration of sediment

2. The sequencing of construction activities to avoid introducing contaminants into building finishes or finished areas is an example of how green projects _____.
   a. maintain construction schedules
   b. reduce soil contamination
   c. protect construction workers
   d. protect future building occupants

3. Choosing construction processes that minimize the use of energy and natural resources is an example of how green projects protect _____.
   a. construction workers
   b. building occupants
   c. future generations
   d. the owner

4. A fence used to protect a mature tree from damage on a site should be placed _____.
   a. close to the trunk of the tree
   b. around the entire area surrounding the tree
   c. at the drip line of the widest branch
   d. at least two feet away from the trunk

5. Projects seeking LEED certification must meet the National Pollutant Discharge Elimination System requirements _____.
   a. only if the project site is larger than one acre
   b. only if part of the development is larger than one acre
   c. only if the development is larger than five acres
   d. always, no matter what size the project is

6. Where possible, site grading should be done to encourage _____.
   a. rapid stormwater flow
   b. erosion flow
   c. sheet flow
   d. concentrated flow

7. Diverting stormwater to vegetated areas can both slow water velocity and _____.
   a. reduce stormwater quantities
   b. reduce infiltration
   c. increase sedimentation
   d. increase sediment filtering

8. The best approach to solving pollution problems on a green construction site is to use solutions that address problems _____.
   a. when they happen
   b. after they happen
   c. before they happen
   d. during construction

9. A proactive prevention measure of pollution prevention of soil and groundwater on a construction site is _____.
   a. a spill cleanup plan
   b. contained storage for hazardous chemicals
   c. spill countermeasures
   d. proper equipment maintenance

10. By using recycled content paper and office products in the on-site construction offices, you help to close the recycling loop by _____.
    a. providing a market for recyclable materials
    b. reducing waste in the construction office
    c. increasing awareness of environmentally friendly products
    d. providing recycling opportunities to employees

Review Questions

11. Temporary site utility lines should be monitored for _____.
    a. leaks
    b. usage
    c. location
    d. repairs

12. The Sustainable Construction Supervisor should ensure that those educated about the project's sustainability goals include all contractor employees on-site, as well as _____.
    a. owners
    b. designers
    c. subcontractors
    d. inspectors

13. Products delivered to a green construction site should seek to minimize transportation costs and _____.
    a. delivery time
    b. packaging requirements
    c. vehicle size
    d. lead time

14. When storing and staging materials on a project site, it is important for indoor air quality to protect the material from damage caused by _____.
    a. mold
    b. moisture
    c. sunlight
    d. rust

15. It is very important to protect plastic products from exposure to _____.
    a. moisture
    b. cold temperatures
    c. UV radiation
    d. hot temperatures

16. A material inventory is important to a materials management plan as it is the basis for identifying _____.
    a. materials on the site
    b. potential threats to materials
    c. schedule impacts
    d. cost implications of green materials

17. A typical cause of damage to installed systems or a need to rework finishes is _____.
    a. careless material management
    b. poor site layout
    c. inefficient procurement of products
    d. poor coordination of subcontractors

18. With respect to sustainability goals, balancing cut-and-fill requirements on a site reduces the _____.
    a. need to dispose of soil spoils offsite
    b. amount of heavy equipment on-site
    c. cost of excavation
    d. need to stockpile topsoil

19. An alternative to building demolition is building _____.
    a. salvage
    b. deconstruction
    c. reconstruction
    d. recycling

20. The effectiveness of composting waste materials on-site is a function of the type of waste generated and the availability of _____.
    a. waste materials
    b. local facilities
    c. labor
    d. compost

21. Incorrect waste sorting on a site can lead to contaminated loads that are rejected by _____.
    a. recyclers
    b. inspectors
    c. quality control officials
    d. regulators

22. For LEED certification, proper tracking documentation of waste quantities and _____ are required for calculations.
    a. type
    b. volume
    c. destinations
    d. costs

Review Questions

23. The construction process that suppresses dust and particulates is the _____.
    a. dry process
    b. wet process
    c. cutting process
    d. gluing process

24. Prior to installation, all duct and equipment openings should be _____.
    a. sealed with plastic
    b. covered with paper
    c. completely dry
    d. checked for leaks

25. A simple method used to prevent construction dirt and grime from contaminating previously installed materials is through _____.
    a. design
    b. scheduling
    c. quality control
    d. inspections

26. In IAQ management planning, it is better to select options that prevent pollution rather than those that _____.
    a. fix it after the fact
    b. make use of VOCs
    c. require inspector approval
    d. require designer approval

27. In pursuing LEED certification, a sustainability team may choose to conduct indoor air quality testing instead of taking time from the schedule for _____.
    a. inspection by USGBC
    b. documentation of VOC emissions
    c. flush-out of the building
    d. review of products used

28. Prior to performance testing, the Sustainable Construction Supervisor will work with the commissioning agents during the _____.
    a. design phase
    b. construction phase
    c. final closeout phase
    d. equipment start-up phase

29. A system performance test occurs _____.
    a. after closeout of the project
    b. before the start-up of a system
    c. after all system components are installed
    d. before all system components are installed

30. In order to meet LEED requirements for SS Prerequisite 1: Construction Activity Pollution Prevention, the ESC plan must include specifications that detail the _____.
    a. schedule of construction
    b. best management practices used
    c. costs associated with erosion control
    d. amount of stormwater on-site

31. In order to meet the LEED requirements for the SS Credit 5.1: Site Development – Protect and Restore Habitat, documentation must be provided that includes a site plan with limits of _____.
    a. excavation indicated
    b. disturbance clearly marked
    c. erosion and sedimentation measures
    d. proposed disturbance after construction

32. The contractor is responsible for ensuring that the commissioning agent has access to all _____.
    a. site plans
    b. design-phase submittals
    c. construction-phase submittals
    d. operations and maintenance manuals

33. The commissioning agent is responsible for preparing an operations manual for each commissioned system; this manual is _____ the contractors operations and maintenance manual provided by the contractor.
    a. the same as
    b. separate from
    c. coordinated with
    d. approved with

Review Questions

34. To achieve the LEED EQ Credit 3: Construction IAQ Management Plan, a project must have an IAQ management plan and documentation of how the plan was _____.
    a. approved
    b. designed
    c. inspected
    d. implemented

35. Documentation for EQ Credit 3.1 requires a detailed _____.
    a. photo log
    b. inspection plan
    c. weather log
    d. cost documentation

36. Building flush-out or baseline IAQ testing occurs after all finishes and furnishings have been installed but before the _____.
    a. punch list is complete
    b. contractor leaves the job site
    c. building is inspected
    d. building is occupied

37. On a LEED project, before making any substitutions or changes in the construction process, a Sustainable Construction Supervisor should _____.
    a. determine the impact on LEED credits
    b. assess the feasibility of the substitution
    c. identify other alternatives
    d. consult with the owner

38. Changes in material installation schedules may negatively impact LEED goals and will require coordination with the _____.
    a. owner
    b. commissioning authority
    c. building inspectors
    d. designers

"The Sustainable Construction Supervisor curriculum provides a much-needed resource for contractors and field associates, who are the most critical link of the sustainable design and construction process, by helping to ensure the work put in place every day has the least impact to the environment possible."

Amanda Atkinson, LEED AP, Senior Manager, Sustainable Services, Holder Construction Company

5.0.0 High Performance Green Project Teams

With the knowledge of so many important features of sustainable projects, Sustainable Construction Supervisors play an important role in leading implementation on site. This section introduces the skills needed to champion sustainability and LEED-related goals of the project to subordinates and other project stakeholders. This section focuses on the roles and responsibilities of project stakeholders in implementing green projects and provides guidance on how to effectively supervise others. It shows how to communicate necessary information to motivate green behavior in others. Specifically, this section provides the knowledge needed for Sustainable Construction Supervisors to achieve the following objectives:

- Understand the roles and responsibilities of project stakeholders in achieving a green project.
- Explain why timely and accurate tracking and verification of project documentation are critical to the project's success in LEED certification.
- Describe why sustainable construction practices benefit not only the project but also the environment and society as a whole.
- Explain the **business case** for sustainability in a way that project stakeholders can understand. Some of the benefits of sustainable design are listed in *Appendix B*.
- Participate as an effective member of a larger green project team.

5.1.0 Implementing a Green Project: Roles and Responsibilities

Many people and organizations participate in the life cycle of a building. *Figure 87* shows the various stakeholders involved throughout the life cycle. This figure shows the important roles that each stakeholder plays. How they influence the success of a green project depends upon their involvement in particular stages of the building life cycle. All stakeholders are critical to the success of a green project.

External stakeholders are based external to the facility, and include contractors, designers, government agencies, and others. For these stakeholders, the built facility represents one of many projects in which they may be involved at any given time.

For internal stakeholders, the building represents a major interest in which they are vested, and it may be the only building affecting them at any point in time. These stakeholders, such as owners, tenants, users, and clients, have a direct stake and involvement in the facility and the functions it serves. It is their needs the facility is designed and constructed to meet.

Direct stakeholders are those entities whose actions directly affect the facility system. They may also be directly impacted by the behavior of the facility. Their needs are met directly by their interaction with the facility. These stakeholders may be either internal or external to the system. Direct stakeholders include users, constructors, designers, owners, and surrounding communities.

Indirect stakeholders have no direct impacts on the facility. They may have no direct interaction with the facility at all. However, they are indirectly impacted by the existence of the facility system. Indirect stakeholders include entities who manufacture materials and supplies used to construct the facility, handle waste materials emitted by the facility, invest money in the potential of the facility, and create codes and regulations which must be observed by the facility system.

In some phases of a project, stakeholders who were indirectly represented by other stakeholders in earlier phases become direct stakeholders. This occurs as their participation in the interactions of the system becomes important. For example, future users and tenants of the facility system often do not participate directly in the planning and design process. They are represented by the owner and/or developer of the facility during those phases. After the facility reaches completion, these groups become direct stakeholders. This is because of their direct participation in the system operation.

5.1.1 Pre-Construction Stakeholders

Pre-construction stakeholders are involved during the planning, pre-design and design phases of the building life cycle. The planners, developers, and owners play a direct role in determining the green goals for a project. They define the approach and set the vision for the design and construction of the facility. It is their responsibility to communicate these goals to the designers. They must also ensure that their ideas for the building align with those of the investor. The ability of a Sustainable Construction Supervisor to influence the design at this stage may be limited. It may only involve input relating to the constructability of the project. Whenever possible, processes that will streamline construction, minimize waste, and influence energy efficiency should be identified.

Indirectly, regulatory agencies can play a large role in the increase of the numbers of green buildings. This can be accomplished by providing developers with financial and regulatory incentives to build green.

Phase	External Stakeholders – Indirect	External Stakeholders – Direct	Internal Stakeholders – Direct	Internal Stakeholders – Indirect
PLANNING/ PRE-DESIGN	• ZONING AGENCIES • REGULATORY AGENCIES	• PLANNERS • DEVELOPERS	• OWNERS • LAND DEVELOPERS	• INVESTORS
DESIGN	• FINANCIERS • CODE ENFORCEMENT AGENCIES • MANUFACTURERS • PROFESSIONAL INSTITUTIONS • SURROUNDING COMMUNITIES	• DESIGN TEAM: – ARCHITECTS – ENGINEERS – INTERIOR DESIGNERS – LANDSCAPE ARCHITECTS • PROJECT MANAGERS	• OWNERS • LAND DEVELOPERS	• INVESTORS • USERS/TENANTS • FACILITY MANAGERS/ OPERATORS • CLIENTS/PRODUCT CONSUMERS
CONSTRUCTION	• MANUFACTURERS • VENDORS/SUPPLIERS • SHIPPERS • CODE ENFORCEMENT AGENCIES • REGULATORY AGENCIES	• CONSTRUCTION TEAM: – CONTRACTORS – CONSULTANTS – SUBS • UTILITIES • FINANCIERS • PROJECT MANAGERS • SURROUNDING COMMUNITIES	• OWNERS	• INVESTORS • USERS/TENANTS • FACILITY MANAGERS/ OPERATORS • CLIENTS • PRODUCT CONSUMERS
OPERATIONS & MAINTENANCE	• MANUFACTURERS • VENDORS/SUPPLIERS • SHIPPERS • REGULATORY AGENCIES	• SURROUNDING COMMUNITIES • UTILITIES • FINANCIERS	• OWNERS • USERS/TENANTS • FACILITY MANAGERS • OPERATORS • CLIENTS	• INVESTORS • PRODUCT CONSUMERS • USER'S DEPENDENTS
DECONSTRUCTION/ DEMOLITION/ DISPOSAL	• WASTE DISPOSAL COMPANIES • RECYCLING COMPANIES • REGULATORY AGENCIES	• SURROUNDING COMMUNITIES • DEMOLITION CONTRACTOR • DISPOSAL AGENT • SALVAGE AGENTS • DEVELOPERS	• OWNERS	• FUTURE USERS • INVESTORS

201F87.EPS

Figure 87 Stakeholders over the facility life cycle.

Green projects are most successful when the project delivery process is integrated. The design team's role is to incorporate sustainable features to meet the goals of the owner and other stakeholders. This is also a phase to include the construction team. Integration of these stakeholders can lead to both first-cost savings and resource savings in other phases of the building life cycle.

5.1.2 Construction Stakeholders

The major role of the construction team is to build the design and track, manage, and verify project documentation required to achieve green goals. The role of the Sustainable Construction Supervisor is to ensure that all stakeholders are participating to the required level. This includes the performance of contractors, subcontractors, suppliers and vendors. The Sustainable Construction Supervisor can exert the most influence during the construction phase. During this phase, the Sustainable Construction Supervisor can encourage workers, subcontractors, and suppliers to identify innovative practices. It is also the critical phase where LEED certification can be jeopardized if all procedures are not followed. It is the responsibility of the Sustainable Construction Supervisor to ensure that all stakeholders are aware of the requirements.

Successful LEED projects require the collection of project documentation. It is also important to accurately track these documents. The Sustainable Construction Supervisor must ensure that all products and materials used on site are within the acceptable green practices. The Sustainable Construction Supervisor must also be able to communicate the importance of the documentation to all stakeholders.

5.1.3 Post-Construction Stakeholders

The role of owners, operators, and facility managers is to operate and maintain the building according to the goals and guidelines developed during the pre-construction phase. They are responsible for the monitoring of systems within the facility to ensure they are working to the desired level. At this stage of the project, facility occupants move into the building and begin to enjoy the benefits of green building.

5.2.0 Communicating the Costs and Benefits of Sustainability

There are many benefits associated with building sustainably. However, some owners may feel that the first costs associated with building green are too high. While this may be true of some elements of environmentally responsible building, it is only for the short term. Several studies have shown that the long-term benefits associated with green buildings far outweigh the initial costs. Even if the first costs are higher, they can be recovered. All stakeholders involved in the development and construction of a green building can benefit. The major challenge is to communicate these benefits to the stakeholders. This can be accomplished by determining what motivates each stakeholder. This information can then be used to develop a strategy that ties the benefits of green building to the issues they consider important.

5.2.1 The Owner's Perspective

Many owners base their decisions for either building green or not building green on first cost. The cost of building sustainably can be justified on several levels. These include both quantifiable costs and qualitative costs. Many green design strategies and construction practices reduce construction costs. However, if owners limit themselves to only these strategies, they may miss the opportunity to save money later in the life of the building.

Often, green buildings cost no more to construct than conventional buildings. This is due to resource-efficient strategies that allow for a downsizing of costly systems. These include mechanical, electrical, and structural systems. Because sustainable buildings are designed to be energy-efficient buildings, the savings in energy costs throughout the life cycle are substantial.

Green buildings also provide high-quality work environments. Many of the energy-efficient components of green buildings also enhance worker performance. Increased daylighting, task lighting, fresh air, and personal environmental controls result in improved employee productivity and reduced absenteeism. Implementing green strategies can result in increases in occupant performance measures by 6 to 26 percent. These strategies also improve the health and well being of occupants. Green buildings offer a healthier indoor environment. This also helps to reduce lost work time due to respiratory diseases and allergies. Healthy buildings also reduce potential legal claims for an owner. Sick-building cases have increased over the years. Some industry experts are even predicting that insurance companies will start linking lower premiums to high performance buildings.

For owners of green buildings, the features associated with these facilities attract and retain tenants. These include features designed to reduce water and energy consumption. Because green buildings maintain their value through higher occupancy and easier facility maintenance, they are easier to sell. They also command a higher market valuation.

When considering whether or not to build green, owners can also consider financial and regulatory incentive programs. Many of these programs are location dependent, but they offer significant advantages to those who choose to build green.

In addition to the financial benefits associated with building green, these facilities also benefit the community. Building green can further define an owner's commitment to the environment, employees, and the community.

The owner is the stakeholder who determines the goals for a green building project. Owners establish the vision that the designers and constructors produce and the facility managers maintain. Achieving this goal requires the cooperation of all parties involved. As building green gains momentum, it is not only a challenge to convince owners of the value, but it will also be important to mentor other project team members.

5.2.2 Mentoring and Managing Subcontractors and Subordinates

Green technologies require new ways of thinking. The most successful green projects involve all stakeholders during the initial stages of design and planning. This early involvement of subcontractors in the design process is a way to help streamline the construction process. Subcontractors provide valuable advice regarding improvements in sitework, building systems (including electrical, mechanical, plumbing, and structural), and the building envelope.

Developing training programs that encourage the participation of subcontractors in the design process is one way to mentor. Subcontractors may be unwilling to participate in the design process. They may feel that they are placing themselves at a disadvantage by revealing their efficient ways of performing work. Communicating the long-term advantages and benefits associated with an integrated design approach to subcontractors is important. These include improved design documents, fewer change orders and potential repeat customers.

Subordinates must also understand the benefits of building sustainably. Education and training programs are a good place to start. These programs should be designed to communicate the benefits of sustainability. This also establishes a firm commitment to the fundamentals of green construction practices. Employees must also understand how their efforts will help to meet the goals of the company and the project. Clear and open communication to subordinates will reduce the risk of confusion that could possibly affect the achievement of sustainability goals. Creating a clear channel of communication will also help to encourage subordinates to initiate innovative solutions to sustainability issues. It will also empower subordinates to raise concerns regarding sustainable practices to superiors.

5.3.0 Identifying and Overcoming Barriers to Sustainable Practices

Sometimes, barriers get in the way of implementing sustainability practices. These barriers are related to individual perceptions, resources, and unclear communication of sustainable goals. The key to moving beyond these barriers involves three steps: evaluating the situation, identifying the obstacles, and developing best practices.

5.3.1 Evaluating the Situation

In order to determine the best course of action to implement green practices, one must first understand the existing situation. Determining the readiness of a group to change is critical. Also important is determining the nature of any resistance to change. These factors will shape the approach to implementing new processes. This process involves assessment of current attitudes and experiences related to sustainability. It will also require an assessment of the willingness to commit resources to the effort.

A good place to begin is to evaluate whether the organization is aligned with sustainable values. Are employees encouraged to suggest ideas? Are sustainability principles included in the company's vision statement? Does the organization feel that social and environmental impacts are also important? Asking these questions will help to determine if the organization believes in a sustainable approach.

The next level of evaluation involves identifying the motivation for embracing sustainability principles. Are the motivations driven by profits, regulations, or the competition? These motivations will determine how to frame new approaches. The ability of an organization to change must also be assessed. Is the organization ready to change? Do they believe that change is necessary? Is the organization ready to devote the required resources to encourage sustainability and sustainable practices?

Upon completion of the evaluation process, the next step is to identify the potential obstacles that may be encountered when trying to implement sustainable practices.

5.3.2 Identifying Potential Obstacles

Change in organizations tends to happen slowly. This pace must quicken in the construction industry if contractors are going to remain successful. The ability to implement a sustainable construction project that is seeking LEED certification is a qualification that owners will be demanding. The adoption of sustainable principles as part of their mission is critical to success and competitiveness in the market.

A Sustainable Construction Supervisor can facilitate the acceptance of sustainable practices by helping individuals accept the need for change. Understanding that the organization may lack the necessary knowledge to adopt green practices is a great place to start. Making the case can be as simple as educating them about successful green projects for similar jobs. By highlighting projects that are realistic, management and subordinates may begin to see ways in which they can adopt green practices. As all members of the organization begin to understand the benefits of green practices, the concepts will begin to be accepted as the standard way to conduct business. To achieve this acceptance, processes must be developed that address their fears of adoption.

5.3.3 Developing Best Practices

One of the most common obstacles to implementation involves the perception that sustainable projects cost more up front. To help the organization see the benefits of sustainable practices, consider beginning with low-cost or no-cost solutions to establish a history of successful practices. It is also helpful to identify free external resources or cost offsets to gain support.

If the organization is resistant to change, the most likely root cause is a lack of motivation to change or a lack of knowledge regarding sustainable practices. This also leads to a lack of management buy-in. This is where the Sustainable Construction Supervisor can begin to articulate the costs and benefits associated with green practices. Help to educate management in the organization about the importance of adopting these practices and how it will impact the triple bottom line—profit, society, and the environment. Begin to develop awareness of the practices and get everyone talking about it on the job site.

When the organization accepts that sustainability is not going away and that green practices are good for the bottom line, the change process has slowly begun. However, one of the toughest obstacles to overcome is a failure or bad experience while trying to adopt sustainable practices. A clear understanding of the situation must be achieved to determine the underlying cause of the failure. Work to develop procedures that will ensure the failure does not happen again and assure management and subordinates that the payoffs will be worth the effort to try again.

Another barrier to adoption of sustainable practices is unclear payoffs or measures of success. A Sustainable Construction Supervisor can help to define the processes and metrics that will be used to define success and is the front-line person on the job site, dealing with stakeholders both up and down the chain of command. This includes owners, designers, engineers, subcontractors, vendors, suppliers and your own work force. The influence of a Sustainable Construction Supervisor on project performance is measurable. Begin by helping to develop metrics that are performance based and life cycle-oriented.

Each obstacle identified in the situational evaluation must be addressed. Most can be addressed through clear communication, increased education, improved knowledge, and better ways to evaluate success.

5.4.0 Contributing to the Green Project Team

Green project teams include many different individuals. Creating an environment where all parties work together to achieve green goals is critical to the success of a project. Each member has a role to play. It is the responsibility of supervisors to ensure that crews comprehend their role in a project. Supervisors must also communicate the benefits of green practices. Finally, supervisors must set the example and encourage their crews to identify new ways to be green.

5.4.1 Defining the Role of a Sustainable Construction Supervisor

A Sustainable Construction Supervisor plays a large role in how labor crews implement green practices on a job site. Some of these practices are not new. For example, sedimentation fences have been a part of construction sites for a long time. However, the monitoring and assurance of their performance may not be second nature. As the leader of the work force, the Sustainable Construction Supervisor must educate the crew and identify sustainable practices early in the project.

A Sustainable Construction Supervisor is responsible for discussing the green practices that will be implemented during a project. It is important that individuals understand the practices that they are expected to perform. This includes comprehension of how their actions impact the achievement of particular LEED credits. It is also important for supervisors to inform workers about all the green practices on the project that impact LEED credits, even if these practices are not the direct responsibility of all workers. Awareness of all issues surrounding LEED certification can help ensure a successful project. This information also empowers workers to take ownership of the green practices on the site. It will also improve their buy-in to the concepts of LEED and green practices. Involvement of the entire work force is critical to the success of the project. Each worker is a steward of the environment. It is your responsibility to ensure that they understand their role.

5.4.2 Advocating Green Practices

On the job site, the implementation of green practices will require educating the work force. Crews will implement green practices daily and they must understand how their actions fit into the overall project. This is a difficult task. If not approached correctly, they may lose interest and not buy into the process, which is critical to the success of the project. Changing the minds of individuals involves altering the way in which they view or think about an issue. It also involves changing the way they do work.

It is important to create a sense of urgency around green practices. Workers must understand why there is a need to change their current practices. The global perspective of reducing the impact that construction has on the environment is a good place to start. Begin with facts about the impacts of construction. Describe how buildings deplete resources, contribute to global climate change and increase greenhouse gases. These are global issues that appeal to a person's sense of doing the right thing. However, this is not enough. An argument must be made that identifies how their specific actions on a project will impact the environment, the project success, and the potential for future work.

It may be necessary to frame an argument for green practices in different ways to appeal to different people. Many green practices require a change in how things are done on a job site. For example, construction waste management plans are becoming more common in the construction community. These plans require the development of procedures. They also require training of the work force. The justification for implementing the plan may have to be explained in several ways. For instance, some workers may feel that separating construction waste is time consuming and not worth their effort. Emphasize the importance of this simple action. Explain how this simple step will help to achieve a LEED credit. While this may appeal to some workers, not all may feel that a LEED credit is a good reason. If this is the case, try framing the argument in terms of the impact this action has on the environment. This would involve highlighting how separating waste and recycling materials helps to reduce the amount of waste that ends up in landfills.

The achievement of project LEED goals and reduced landfill wastes are good reasons to implement a construction waste management plan. However, these reasons may not appeal to all workers. Another approach is to identify how these practices will impact the potential for future work. Construction companies that implement these plans may be more attractive to owners during the bidding process. Also, if the company is familiar with the procedures required to implement a waste management plan, they may be able to submit a more competitive bid package. Finally, workers who comprehend green practices will be more valuable in the work force.

Continuing with the construction waste management plan example, another way to create enthusiasm for green practices is involvement of workers in the development of the plan. Including the individuals who will have to follow the procedures developed during the planning stages is an easy step that will achieve many benefits in the future. Having participated in the planning, the workers know why they have to complete certain tasks. They can trace their individual contribution to the overall project sustainability goals and the achievement of LEED points. Additionally, this knowledge will also increase their marketability as green workers.

Overall, it is important to know what motivates the work force. Being able to communicate benefits of green practices requires supervisors to appeal to many different audiences.

5.4.3 Setting the Example

The example that must be set on a green project occurs at several levels. First, employees must understand the commitment that the owner has to achieving a green building. This may be demonstrated to the entire project team through an integrated design process, informational sessions highlighting the green components of the project, educational flyers, or other programs that highlight their commitment.

Secondly, workers must understand the commitment of their company to green practices. Are green practices employed only when working on a project site that requires them, or does the company embrace these concepts in their daily practice? Company commitment can be demonstrated through several mechanisms. Some of these include the development of company green policies, training employees in fundamental green technologies, accrediting supervisors in one of the green accreditation programs, and providing incentives for employees who embrace green concepts.

Providing opportunities on the job site for employees to behave more sustainably is also a way to set an example. Simple efforts like providing recycling bins and encouraging car-pooling can go a long way. More complex efforts, like purchasing hybrid vehicles for use on site, are additional ways to set examples.

Finally, and most importantly, the right example must be set in the field. The Sustainable Construction Supervisor interacts daily with the individuals who implement the green practices. The supervisor's actions set the tone for the entire project. If these actions do not match the supervisor's expectations for the crew, it can derail the entire effort. It is important that workers observe supervisors acting in a sustainable manner. Practicing green habits is one way to set the example, both on and off the job site.

Simple steps such as recycling beverage containers, newspapers, and other recyclable materials in the office trailer can illustrate the level of commitment. Additionally, be aware of other green practices that can influence the buy-in of the crew. For example, the commitment to green practices may be questioned if a supervisor drives a gas guzzler rather than a fuel-efficient vehicle.

Personal actions are not the only way that to set an example. Holding others accountable to green practices is important. Consistent enforcement of green practices is one way to ensure that workers begin to adopt these practices. This requires daily observation and encouragement of workers who are achieving the goals of the project. The overall attitude of a crew can be affected by the actions of the supervisor, so be sure to send a consistent message to the work force.

5.5.0 Identifying Opportunities for Field Innovations

Many innovations are developed in the field to improve performance or streamline processes. Some of these innovations may be applicable to the achievement of an innovation credit under the LEED rating system. This credit is designed to provide project teams the opportunity to be awarded points for exceptional performance above the requirements set by LEED. The credit may also be achieved by an innovative performance in a green building practice that is not specifically addressed by LEED. To achieve these credits, the team must document the achievement by the LEED credit equivalence process.

To encourage workers to identify these opportunities, Sustainable Construction Supervisors should develop open lines of communication. By establishing trust early in the project, subordinates are more likely to communicate field innovations. Another approach is to review the design documents and specifications with crews and specifically ask them how they would accomplish a task more efficiently. Asking them for their thoughts and ideas is another way to encourage participation.

Another method that has proven successful for encouraging innovation input is an incentive program. The program would reward an individual or crew that developed an idea for an innovation credit. The incentive could be monetary or time off with pay, or simply recognition in a company newsletter. Construction crews have a vast amount of experience and their knowledge can lead to innovative sustainable solutions.

"It has been a pleasure working on the NCCER Sustainable Construction Supervisor program that will extend sustainability to the field supervisors and educate them on the best practices to implement. This will only help in increasing awareness of sustainable practices to those who are responsible for field operations."

JJ McCarthy , LEED AP, Janus Development Services LLC

Review Questions

1. Contractors are what type of stakeholder in a green project?
   a. Remote
   b. Indirect
   c. Internal
   d. External

2. Many owners base their decisions about whether to build green or not to build green upon _____.
   a. life cycle costs
   b. maintenance costs
   c. green enhancement costs
   d. first costs

3. A major advantage of green buildings is the increase in worker _____.
   a. interest
   b. productivity
   c. attitudes
   d. absenteeism

4. Green buildings are easier to sell because they maintain their value through higher occupancy rates and _____.
   a. increased energy costs
   b. increased water usage
   c. easier facility maintenance
   d. easier tenant improvements

5. An easy step to help encourage subcontractors to participate in green projects and to contribute during the design phase is to develop _____.
   a. an incentive program
   b. a training program
   c. new relationship with outside subcontractors
   d. improved maintenance requirements

SUMMARY

Traditional construction practices have significant short- and long-term impacts on the natural and human environment. To reduce negative impacts, sustainable construction practices are used to plan, design, and build construction projects. Managing a project to increase its sustainability requires careful attention to construction operations. Both the project sustainability team and the Sustainable Construction Supervisor make decisions that are important for project sustainability.

Critical issues affecting project sustainability include damage to the environment, use of resources, and generation of waste. Environmental damage to a project site during construction affects a project's sustainability and can impact neighboring sites. The products and technologies used to build the project have effects on sustainability over their entire life cycle. Construction waste is one of the most apparent impacts of the construction process. Reducing or recycling construction waste reduces demand for virgin materials and reduces the negative impacts of raw material production and product disposal.

The Sustainable Construction Supervisor plays an important role in helping a project meet sustainability goals along with the requirements of the LEED-New Construction rating system, Version 3 (2009). Prior to construction, the design team works on design-phase credits and documents them for design review. After the design-phase LEED review is complete, it is the construction team's responsibility to ensure that these credits are implemented as designed. The construction team is also responsible for construction phase credits targeted for the project.

In addition to supporting LEED certification, the specific responsibilities of the Sustainable Construction Supervisor will vary from project to project and company to company. When procuring materials for the project, the Sustainable Construction Supervisor should be aware of which products have been targeted to meet sustainability goals by the project sustainability team. A critical role is to evaluate the effect on LEED credits when subcontractors propose product substitutions. Depending on the project delivery method being used, the Sustainable Construction Supervisor may be involved in commissioning activities throughout the project delivery process. The Sustainable Construction Supervisor is also responsible to ensure that requirements are met for indoor air quality management, material storage, and verification of correct product installation.

For each project, the Sustainable Construction Supervisor should carefully review project sustainability plans and all construction documents to identify specific contractor requirements. Construction-phase credits depend on proper documentation to meet certification requirements. Green projects are most successful when the project delivery process is integrated. The Sustainable Construction Supervisor is an important player who leads activities on the project site and works with the project team to create the most sustainable project possible.

Trade Terms Introduced in This Module

Agricultural land: Property suitable for raising crops or livestock. Includes land cultivated for crops, orchards or vineyards, as well as meadows and pastures used for grazing.

Airborne dust: Particles of soil or other materials that are small enough to be suspended in the air and breathed by organisms.

Albedo: The fraction of solar energy reflected back after striking a surface.

American National Standards Institute (ANSI): An organization that sets standards for product performance and testing.

American Society for Testing and Materials (ASTM) International: An organization that sets standards for material performance and testing.

Best practice: An action or technology to achieve a desired end that is agreed upon by industry to represent the most effective or efficient way to achieve that end.

Biodiversity: The number and variety of organisms found within a particular geographic region.

Bioretention cells: Engineered areas designed to capture and treat stormwater runoff and remove impurities before it percolates into the soil.

Blast furnace slag: The impurities left over during the process of smelting ore.

Brownfield sites: Areas of land that are contaminated or perceived to be contaminated by some past industrial activity.

Building envelope: The elements and systems of a building that separate the interior of the building from the outside; it includes the roof, walls, windows, and doors.

Business case: The set of pros and cons, often expressed in economic terms, associated with a proposed action that defines how that action will contribute to a business's enterprise.

Certified sustainable harvest: The extraction or collection of a natural product in a way that can be continued indefinitely without damage, observed and verified by a third party.

Certified wood: Wood or wood products that have been verified as coming from sustainably harvested sources by a third-party agency.

Commissioning: A process of quality assurance undertaken during project delivery to ensure that all building systems function according to their design intent.

Composting: Purposeful biodegradation of organic matter, such as yard and food waste, into a soil amendment to improve soil quality.

Corporate social responsibility (CSR): The practice of managing or operating a company to have an overall positive impact on society.

Cradle2Cradle: A rating and evaluation system for products, developed by the McDonough Braungart Design Collaborative, that evaluates a product's impacts from its inception through recovery and reuse at the end of its life cycle.

CRI Green Label: A rating system developed by the Carpet and Rug Institute to evaluate the indoor air quality performance of carpet and flooring products.

Deconstruction: The careful disassembly of a built facility in a way that enables components to be reused on future projects.

Downstream impacts: Effects occurring as a result of the use of a product, or as a result of its disposal at the end of its service life.

Eco-label: A mark or certifying logo that indicates an environmentally-friendly attribute of a product.

Ecological rucksack: The total quantity of materials moved or used to create a product or service.

Ecosystem: An ecological community together with its environment, functioning as a unit.

Embodied energy: The total energy that was used in all the activities associated with producing a product and delivering it to its point of use.

Emissions: Substances discharged into the air or water as pollutants.

End-of-life cycle: The point in time when a product is no longer suitable to be used for its intended purpose.

Energy performance credits: A series of credits in the Energy and Atmosphere category of the LEED rating system that reward projects for exceeding the energy code by a certain amount.

Energy Star: A rating and benchmarking system developed by the U.S. Environmental Protection Agency to evaluate the energy performance of buildings and products.

Enhanced commissioning: A process of quality assurance that begins during design and continues throughout and following construction to ensure that all building systems function according to their design intent.

Environmental stewardship: Responsibility for environmental quality that is shared by those whose actions affect the environment.

Erosion: The action of wearing down an area of exposed soil due to air or water running over it.

Erosion and Sedimentation Control (ESC) plan: A plan developed for a project that identifies risks for erosion and sedimentation during the project and specifies measures for preventing or mitigating them.

Externalities: Consequences of an action that are experienced by a third party not participating in the action.

Fitness for purpose: The ability of a technology or product to fully meet the intent for which it was designed.

Flush-out: The process of operating a building's mechanical ventilation systems at full volume with outside air to remove air pollutants from a building prior to occupancy.

Footprint: Refers to the effect of human activity on the natural environment; for example, the ecological footprint or building footprint.

Forest Stewardship Council (FSC) Certified: Wood or wood products that have been certified by the Forest Stewardship Council.

Fossil fuels: Hydrocarbon deposits, such as petroleum, coal, or natural gas, derived from living matter of a previous geologic time and burned to produce useful work.

Green Seal: A non-profit, third-party organization that develops standards for evaluating the environmental qualities of products such as paints and adhesives.

Green space: The undeveloped areas around built facilities that are covered in landscaping or vegetation.

Greenfield site: A project site that has not been previously developed except for agricultural use.

GREENGUARD: A rating and certification system used to evaluate the indoor air quality performance of building products and systems.

Greenwash: Dissemination of misleading information by an organization to conceal poor environmental practices and present a positive public image.

Indoor Air Quality (IAQ) Management plan: A plan developed for a construction project that inventories possible threats to future indoor air quality arising from the construction process or materials used in a project, then identifies strategies to mitigate or avoid them.

Infiltration trench: Excavated areas filled with plants and vegetation that are designed to allow stormwater to percolate back into the soil.

Innovation credit: A credit under the LEED rating system awarded for actions that greatly exceed the credit requirements defined under LEED.

International Standards Organization (ISO): A collection of national standards organizations that coordinates international standards.

Just-in-time delivery: The practice of arranging for materials to be delivered to the project site just before they are installed, so that they are not at risk from being damaged during storage.

Leadership in Energy and Environmental Design (LEED): A rating system for green buildings developed by the U.S. Green Building Council.

LEED Online: The website used by members of a project team to manage the process of applying for LEED certification.

Life cycle: The period of time between when a product's raw materials are harvested and when it is ultimately disposed of or recycled, including manufacture and use.

Life cycle assessment: A method that identifies and quantifies the environmental impacts associated with a product over its life cycle.

Life cycle costing: The process of identifying and tabulating all the costs associated with a product or process over its whole life cycle, including first costs, operating and maintenance costs, and end of life cycle costs.

Light pollution: Excessive or unwanted light from artificial sources.

Liquid waste: Materials discharged from a process or entity in liquid form as a pollutant.

Low-emitting substance: A material that emits fewer volatile organic compounds than conventional materials of the same type.

Materials management plan: A plan developed for a construction project that inventories materials involved in the project and identifies strategies to protect them until they can be installed, thereby preventing waste.

Maximum Efficiency Reporting Value (MERV): A means of measuring a filter's ability to remove particles from the air. Ranges from 1 (least efficient) to 16 (most efficient).

Noise pollution: Environmental sound that is annoying, distracting, or potentially harmful.

National Pollutant Discharge Elimination System (NPDES): A part of the Clean Water Act requiring entities to obtain permits specifying the types and amounts of pollutants that may be discharged into water bodies.

National Sanitation Foundation (NSF): An organization that evaluates and certifies plumbing components and other products.

Offgassing: The release of volatile organic compounds from a product or material under normal environmental temperature and humidity conditions.

Particulate: A small mass of solid or liquid matter that remains suspended in a gas or liquid.

Pollution prevention plan: A plan developed for a construction project that inventories potential sources of pollution or waste during construction and identifies strategies to keep them from happening.

Post-consumer recycled content: Waste that comes from a product or material after it has served the useful purpose for which it was made.

Pre-consumer recycled content: Waste that comes from the production process for a product or material.

Rapidly renewable materials: Materials derived from plants that can be regrown over a period of ten years or less.

Recyclability: The capacity of a product or material to be recycled into the same or other products after its service life.

Recycled content: The portion of a material or product that is derived from materials taken from the solid waste stream and reprocessed for use in manufacturing.

Recycled materials: Materials that have been taken from the waste stream and reprocessed for further use.

Regenerative construction: Built environment projects that improve or enhance the ability of the natural environment to function.

Regionality: The degree to which a material or product is produced locally or within the local region.

Salvage: Property or goods saved from damage or disposal for reuse in other applications.

Scientific Certification Systems (SCS): A third-party organization that certifies environmental claims.

Sediment: Small particles that settle to the bottom of a liquid, such as soil particles in a body of water.

Service life: The length of time during which a technology or product can be used for its intended purpose.

Site disturbance plan: A plan that indicates the types and limits of disturbance to be allowed during construction on a project site, and identifies strategies for ensuring that disturbance remains within those limits.

Soil compaction: The compression of soil due to heavy weight or pressure that causes damage to plants and makes it difficult for vegetation to grow.

Solid waste: Materials discharged from a process or entity in solid (not liquid or airborne) form as a pollutant.

Source control: Deliberate actions taken to prevent waste at the point in a process where pollution or waste is generated.

Stormwater runoff: Water from rainfall or snow melt that flows over the ground, especially over impervious surfaces, such as pavement.

Temporary utilities: Services, such as electrical power, water, and communications, that are known in advance to be of limited duration.

Triple bottom line: Accounting for and reporting on not only the financial profits and losses of a company but also the impacts of that company on the environment and society.

Upstream impacts: All of the side effects a product has before it is actually used.

Urban heat island effect: The net rise in temperature in an area resulting from greater absorption of heat by the buildings and paved surfaces in the area.

Vegetated swales: Constructed watercourses covered in appropriate vegetation to slow and retain stormwater.

Virgin materials: Raw materials extracted from nature and used to make products that have not previously been used.

Volatile organic compounds (VOCs): Organic chemical compounds that are vaporous under normal environmental temperatures and humidity conditions. Many of these compounds cause adverse health effects when inhaled.

Waste diversion: Preventing solid waste from going to a landfill and directing it instead to recycling, salvage, or other beneficial use.

WaterSense: A rating and certification system developed by the U.S. Environmental Protection Agency to evaluate the water efficiency of building products.

Wetland: An area of land in which the soil is saturated with moisture either permanently or seasonally and supports plants adapted for life in wet conditions.

Wildlife habitat (habitat): An environment where an organism or community of organisms typically lives or is found.

Appendix A

LEED for New Construction and Major Renovation, Version 3 Registered Project Checklist

SUSTAINABLE SITES — 26 POINTS

Prereq 1	**Construction Activity Pollution Prevention**	Required
Credit 1	**Site Selection**	1
Credit 2	**Development Density and Community Connectivity**	5
Credit 3	**Brownfield Redevelopment**	1
Credit 4.1	**Alternative Transportation**, Public Transportation Access	6
Credit 4.2	**Alternative Transportation**, Bicycle Storage and Changing Rooms	1
Credit 4.3	**Alternative Transportation**, Low-Emitting and Fuel-Efficient Vehicles	3
Credit 4.4	**Alternative Transportation**, Parking Capacity	2
Credit 5.1	**Site Development**, Protect or Restore Habitat	1
Credit 5.2	**Site Development**, Maximize Open Space	1
Credit 6.1	**Stormwater Design**, Quantity Control	1
Credit 6.2	**Stormwater Design**, Quality Control	1
Credit 7.1	**Heat Island Effect**, Nonroof	1
Credit 7.2	**Heat Island Effect**, Roof	1
Credit 8	**Light Pollution Reduction**	1

WATER EFFICIENCY — 10 POINTS

Prereq 1	**Water Use Reduction**	Required
Credit 1	**Water Efficient Landscaping**	2 to 4
	Reduce by 50%	2
	No Potable Use or Irrigation	4
Credit 2	**Innovative Wastewater Technologies**	2
Credit 3	**Water Use Reduction**	2 to 4
	30% Reduction	2
	35% Reduction	3
	40% Reduction	4

ENERGY & ATMOSPHERE — 35 POINTS

Prereq 1	**Fundamental Commissioning of the Building Energy Systems**	Required
Prereq 2	**Minimum Energy Performance**	Required
Prereq 3	**Fundamental Refrigerant Management**	Required
Credit 1	**Optimize Energy Performance** (Number of Credits Based on %)	1 to 19
Credit 2	**On-Site Renewable Energy** (Number of Credits Based on %)	1 to 7
Credit 3	Enhanced Commissioning	2
Credit 4	**Enhanced Refrigerant Management**	2
Credit 5	**Measurement & Verification**	3
Credit 6	**Green Power**	2

MATERIALS & RESOURCES — 14 POINTS

Prereq 1	**Storage & Collection of Recyclables**	Required
Credit 1.1	**Building Reuse**	1 to 3
	Maintain 55% of Existing Walls, Floors, and Roof	1
	Maintain 75% of Existing Walls, Floors, and Roof	2
	Maintain 95% of Existing Walls, Floors, and Roof	3
Credit 1.2	**Building Reuse**, Maintain Interior Nonstructural Elements	1
Credit 2	**Construction Waste Management**	1 to 2
	50% Recycled or Salvaged	1
	75% Recycled or Salvaged	2
Credit 3	**Materials Reuse**	1 to 2
	Reuse 5%	1
	Reuse 10%	2

MATERIALS & RESOURCES (Continued)		14 POINTS
Credit 4	**Recycled Content**	1 to 2
	10% of Content	1
	20% of Content	2
Credit 5	**Regional Materials**	1 to 2
	10% of Materials	1
	20% of Materials	2
Credit 6	**Rapidly Renewable Materials**	1
Credit 7	**Certified Wood**	1

INDOOR ENVIRONMENTAL QUALITY		15 POINTS
Prereq 1	**Minimum Indoor Air Quality Performance**	Required
Prereq 2	**Environmental Tobacco Smoke (ETS) Control**	Required
Credit 1	**Outdoor Air Delivery Monitoring**	1
Credit 2	**Increased Ventilation**	1
Credit 3.1	**Construction Indoor Air Quality Management Plan**, During Construction	1
Credit 3.2	**Construction Indoor Air Quality Management Plan**, Before Occupancy	1
Credit 4.1	**Low-Emitting Materials**, Adhesives and Sealants	1
Credit 4.2	**Low-Emitting Materials**, Paints and Coatings	1
Credit 4.3	**Low-Emitting Materials**, Flooring Systems	1
Credit 4.4	**Low-Emitting Materials**, Composite Wood and Agrifiber Products	1
Credit 5	**Indoor Chemical and Pollutant Source Control**	1
Credit 6.1	**Controllability of Systems**, Lighting	1
Credit 6.2	**Controllability of Systems**, Thermal Comfort	1
Credit 7.1	**Thermal Comfort**, Design	1
Credit 7.2	**Thermal Comfort**, Verification	1
Credit 8.1	**Daylight and Views**, Daylight	1
Credit 8.2	**Daylight and Views**, Views	1

INNOVATION IN DESIGN		6 POINTS
Credit 1	**Innovation in Design**	1 to 5
	Innovation or Exemplary Performance	1
	Innovation or Exemplary Performance	1
	Innovation or Exemplary Performance	1
	Innovation	1
	Innovation	1
Credit 2	**LEED® Accredited Professional**	1

REGIONAL PRIORITY		4 POINTS
Credit 1	**Regional Priority**	1 to 4
	Regionally Defined Credit Achieved	1
	Regionally Defined Credit Achieved	1
	Regionally Defined Credit Achieved	1
	Regionally Defined Credit Achieved	1

PROJECT TOTALS (Certification Estimates) **110 POINTS**

Certified: 40–49 points, **Silver:** 50–59 points, **Gold:** 60–79 points, **Platinum:** 80 points and above

Appendix B

Matrix of Sustainable Design Benefits

The U.S. Department of Energy website at www.eere.energy.gov provides information on the benefits of various sustainable design features (see Table A-1). The features are divided into six primary categories:

1. Sustainable Siting
2. Water Efficiency
3. Energy Efficiency
4. Sustainable Materials and Resources
5. Indoor Environmental Quality
6. Commissioning and Operation & Maintenance (O&M)

These correspond closely to the categories in the Leadership in Energy and Environmental Design (LEED™) rating system and the Whole Building Design Guide available at www.wbdg.org. Thus, this matrix combines the "triple bottom line" benefits framework with six principal elements of sustainable building design. Some of the benefits accrue to the facility owner, and other benefits accrue to other parties such as employees, local governments, the local community/economy, or the public at large. When applicable, those are designated in the table.

Although the list of benefits is extensive, it may not be completely comprehensive. Some of the benefits of sustainable design will be discovered as more experience is gained with living and working in sustainable buildings.

Table B-1 Possible Benefits of Sustainable Design by Feature

SUSTAINABLE DESIGN FEATURE	ECONOMIC BENEFIT	SOCIAL BENEFIT	ENVIRONMENTAL BENEFIT
Sustainable Siting			
Site selection: • Analyze alternatives, including use of existing facilities. • Do not build on prime agricultural land, flood plains, and habitats for threatened species; near wetlands; or on parklands. • Consider urban redevelopment and brownfield redevelopment sites (which require cleanup).	**Facility:** • Lower site preparation costs • Potential for reduced costs of litigation and time delay (e.g., because threatened species are not present at the site) **Local government:** • Lower infrastructure development costs to support existing rather than greenfield sites **Local economy:** • Reduced tourism losses	**Society:** • Preservation of natural areas, agricultural land, and parkland for future generations	**Society:** • Decreased use of virgin resources • Protection of threatened species and wetlands • Remediation of contaminated land (when brownfield is used) • Reduced erosion and flood damage • Reduced impact on fisheries and forests
Site analysis and harmonious building-site relationship: • Inventory and analyze the ecological context, urban and historical context, and natural and cultural attributes. • Organize mass, orientation, topography, and outdoor spaces to employ passive solar principles; provide outdoor spaces; etc. • Design to reduce potentially detrimental conditions, such as slopes that can erode. • Avoid adverse impacts on adjacent properties. • Carefully consider the placement of existing trees on site.	**Facility:** • Possible reduction in first costs (reduced size and cost of mechanical systems) • Reduction in operating costs (fuel costs) **Local government:** • Elimination of unnecessary infrastructure expenditures due to good assessment of site resources	**Society:** • Improved aesthetic and functional quality of site and building for both occupants and neighbors	**Society:** • Reduction in energy consumption and emissions due to optimal orientation, etc. • Conservation and restoration of ecological and cultural resources • Reduced negative microclimate and environmental effects in local vicinity
Facilitation of alternative transportation use: • Locate facility near rail station or bus lines. • Provide covered, wind-sheltered seating or waiting areas for public transport. • Provide bicycle storage and showers. • Install alternative-fuel refueling stations. • Provide preferred parking for carpools.	**Facility:** • Slightly lower capital cost due to reduced parking lot size (could be offset by additional costs for showers and refueling stations) **Employees:** • Potentially lower commuting costs (using public transport rather than personal cars)	**Employees:** • More transportation options	**Society:** • Lower energy use and air pollution from vehicles • Reduced land disturbance for new roads

SUSTAINABLE DESIGN FEATURE	ECONOMIC BENEFIT	SOCIAL BENEFIT	ENVIRONMENTAL BENEFIT
Sustainable Siting, Continued			
Erosion and sedimentation control and stormwater management: • Develop sediment and erosion control plan. • Prevent loss of soil during construction. • Prevent sedimentation of storm sewer or receiving streams. • Plant watershed buffers. • Allow infiltration via porous surfaces. • Filter stormwater through plantings and soil. • Use natural drainage systems.	**Facility:** • Decreased cost of storm drainage construction by using more natural methods • Reduced cost of landscaping after construction is completed (because topsoil is saved) **Local/state government:** • Reduced cost of stream cleanup and water treatment plants	**Neighbors/local community:** • Less disturbance during construction	**Society:** • Less loss of soil during construction • Prevention of sedimentation of storm sewers or receiving streams • Reduced dust/particulate matter during construction • Less disruption of natural water flows • Reduction of runoff into natural water systems
Reduced site disturbance during construction: • Limit site disturbance around the building perimeter, curbs, walkways, and main utility branch trenches. • Reduce the development footprint to exceed open space requirements in local zoning rules	**Facility:** • Potentially decreased cost of clear-cutting and subsequent relandscaping • Decreased site infrastructure costs	**Neighbors/local community:** • Less disturbance during site preparation • Retention of vegetation and scenic vistas • More greenspace for use by occupants and/or community	**Society:** • Preservation of trees and other vegetation • Increased habitat for natural species; biodiversity
Sustainable landscape and exterior design: • Use self-sustaining landscape design and site maintenance procedures. • Restore habitats. • Use plants native to the region. • Consider drought-resistant plants. • Avoid plants needing chemical treatment and fertilizers or causing allergic reactions. • Consider using "green roofs."	**Facility:** • Reduced maintenance costs, water use, fertilizer, and fossil fuel use due to "naturally manicured" landscaping • Possibly decreased heating/ cooling loads due to vegetated roof	**Neighbors/local community:** • More aesthetic natural exterior appearance of building to neighbors and occupants	**Society:** • Reduced impact on microclimate • Lower threat of negative ecological impacts resulting from use of non-native plant species

SUSTAINABLE DESIGN FEATURE	ECONOMIC BENEFIT	SOCIAL BENEFIT	ENVIRONMENTAL BENEFIT
Sustainable Siting, Continued			
Light pollution reduction: • Allow no direct-beam illumination to leave building site. • Use light-colored or reflective edges along driveways and walks to reduce dependence on high-wattage electrical lighting at night. • Use solar lights. • Use security lighting with motion sensors to reduce use of lights at night.	**Facility:** • Possibly lower cost of electricity for lighting (depending on design)	**Neighbors/local community:** • Less disturbance to natural night sky conditions • Improved security of building	**Society:** • Preservation of nocturnal habitat

SUSTAINABLE DESIGN FEATURE	ECONOMIC BENEFIT	SOCIAL BENEFIT	ENVIRONMENTAL BENEFIT
Water Efficiency			
Water use reduction: • Use captured rainwater for landscaping, toilet flushing, etc. • Treat and re-use graywater, excess groundwater, and steam condensate. • Use low-flow fixtures and fittings (pressure-assisted or composting toilets, waterless urinals, etc.) and ozonation for laundry. • Use closed-loop systems and other water reduction technologies for processes.	**Facility:** • Decreased water costs • Potential for O&M savings **Local government:** • Less wastewater treatment infrastructure needed (fewer taxes to pay for infrastructure)	**Society:** • Preservation of natural water resources for future generations	**Society:** • Reduced use of potable water • Reduced generation of wastewater; lower discharge to natural waterways

SUSTAINABLE DESIGN FEATURE	ECONOMIC BENEFIT	SOCIAL BENEFIT	ENVIRONMENTAL BENEFIT
Energy Efficiency			
Space layout: • Fully utilize opportunities for passive solar heating/cooling. • Optimize natural ventilation and daylighting. • Enhance penetration of daylight to interior spaces. • Provide inviting staircases to encourage their use rather than elevators.	**Facility:** • Diminished heating, cooling, and lighting loads and reduced energy costs • Increased operating efficiency due to right sizing of equipment • Possibly lower capital costs due to reduced size/cost of mechanical systems through more efficient design features, appropriate sizing, and optimal integration • Potentially higher occupant productivity due to daylighting **Society/utility companies:** • Avoidance of electricity generation and transmission/ distribution construction costs	**Employees:** • Improved quality of interior space (a secondary benefit of many energy-efficient design features)	**Society:** • Lower electricity use, fossil fuel use, and air pollution/carbon dioxide (CO^2) emissions and other environmental impacts of electricity production and fossil fuel use • Decreased impacts of fuel production and distribution (for fuels used in the building or in production of electricity for the building)
Building envelope: • Design envelope to provide good thermal and moisture control while supporting passive solar and natural ventilation design strategies. • Detail walls to provide best vapor barriers and low outside air infiltration.	**Facility:** • Diminished heating, cooling, and lighting loads and reduced energy costs • Increased operating efficiency due to right sizing of equipment • Possibly lower capital costs due to reduced size/cost of mechanical systems through more efficient design features, appropriate sizing, and optimal integration **Society/utility companies:** • Avoidance of electricity generation and transmission/ distribution construction costs	**Employees:** • Improved quality of interior space (a secondary benefit of many energy-efficient design features)	**Society:** • Lower electricity use, fossil fuel use, and air pollution/carbon dioxide (CO_2) emissions and other environmental impacts of electricity production and fossil fuel use • Decreased impacts of fuel production and distribution (for fuels used in the building or in production of electricity for the building)

SUSTAINABLE DESIGN FEATURE	ECONOMIC BENEFIT	SOCIAL BENEFIT	ENVIRONMENTAL BENEFIT
Energy Efficiency, Continued			
Lighting and sun control: • Use glazing to supply daylighting but control glare. • Use roof monitors and windows set high in the wall (clerestory windows). • Specify photocell-dimming sensors to adjust light. • Use separate switches to turn off lights in individual areas. • Use light shelves and other techniques to bring light deeper into the building. • Supplement daylighting with high-performance lighting that improves visual quality while reducing electrical use. • Use low-ambient lighting levels with task lights, where appropriate. • Use occupancy sensors, dimmers, photocells, and lumen maintenance controls.	**Facility:** • Reduced electricity consumption/costs • Lower cooling loads due to lower heat gains from electrical lighting; hence, reduced energy costs and lower capital costs for cooling system • Potentially higher occupant productivity due to daylighting and visual quality **Society/utility companies:** • Avoidance of electricity generation and transmission/distribution construction costs	**Employees:** • Improved quality of interior space	**Society:** • Lower electricity use and the associated air pollution/CO_2 emissions • Decreased impacts of fossil fuel production and distribution
Systems and equipment: • Optimize mechanical systems to work with the building layout, orientation, envelope, etc. • Consider HVAC zoning, distribution systems, heat recovery systems, modular boilers, and ice storage. • Do not use chlorofluorocarbon (CFC) or hydrochlorofluorocarbon (HCFC) refrigerants. • Develop integrated system designs that consider interaction of systems with overall building layout, envelope, etc. • Include efficient power distribution systems, electrical equipment, motors, transformers, etc. • Consider use of raised floor and underfloor HVAC with personal controls.	**Facility:** • Reduced energy costs • Reduced capital costs due to downsizing of equipment • Lower churn costs (if raised floor is used) • Better productivity (if personal controls are implemented)	**Employees:** • Improved occupant comfort	**Society:** • Lower electricity use and the associated air pollution/CO_2 emissions • Decreased impacts of fossil fuel production and distribution • Protection of the ozone layer (due to avoidance of CFCs/HCFCs)

SUSTAINABLE DESIGN FEATURE	ECONOMIC BENEFIT	SOCIAL BENEFIT	ENVIRONMENTAL BENEFIT
Energy Efficiency, Continued			
Renewable energy: • Consider using photovoltaic, solar heat and hot water, geothermal heat pumps, etc. • Consider entering into green power or renewable credit contracts.	**Facility:** • Lower annual energy costs **Local economy:** • Potential for emerging businesses related to renewable energy	**Society:** • Promotion of market for renewable energy products	**Society:** • Lower electricity use and the associated air pollution/CO_2 emissions • Decreased impacts of fossil fuel production and distribution
Energy load management: • Use energy management systems, monitoring, and controls to continuously calibrate, adjust, and maintain energy-related systems.	**Facility:** • Operational savings (can offset higher first costs) • Reduced capital cost of mechanical systems because control systems reduce the need for oversizing	**Employees:** • Improved comfort, health, and safety	**Society:** • Improved energy efficiency, hence lower electricity and fossil fuel use and lower emissions **Employees:** • Better indoor air quality

SUSTAINABLE DESIGN FEATURE	ECONOMIC BENEFIT	SOCIAL BENEFIT	ENVIRONMENTAL BENEFIT
Sustainable Materials and Resources			
Storage/collection of recyclables: • Provide a system for collecting and storing materials such as paper, glass, plastic, and metals for recycling.	**Facility:** • Possibly lower waste disposal costs **Local community:** • Local recycling business opportunities **Local government:** • Lower landfill construction costs	**Employees:** • Opportunity for building occupants to feel they are "making a difference," which can be an opening for other actions **Society:** • Expanded market for recycled and environmentally preferable products	**Society:** • Reduced strain on landfills • Reduced use of virgin natural resources
Building and resource re-use: • Reuse building shell from existing buildings and fixtures from demolished buildings. • Use salvaged/refurbished materials.	**Facility:** • Decreased first costs due to re-use of materials **Local government:** • Less waste disposal; lower need for new waste disposal facilities	**Society:** • Expanded market for salvaged materials	**Society:** • Reduced strain on landfills • Reduced use of virgin natural resources
Construction waste management: • Recycle or productively use construction, demolition, and land clearing wastes. • Divert these wastes from landfill disposal.	**Facility:** • Possible decrease in construction first costs due to lower waste disposal costs	**Society:** • Expanded market for recycled and environmentally preferable products	**Society:** • Reduced strain on landfills • Reduced use of virgin natural resources
Recycled content: • Select environmentally preferable materials that include recycled materials (use standards developed by government agencies or other reliable sources).	**Facility:** • Possible decrease in first costs of construction due to lower price of recycled materials (in some cases)	**Society:** • Expanded market for recycled and environmentally preferable products	**Society:** • Reduced strain on landfills • Reduced use of virgin natural resources

SUSTAINABLE DESIGN FEATURE	ECONOMIC BENEFIT	SOCIAL BENEFIT	ENVIRONMENTAL BENEFIT
Sustainable Materials and Resources, Continued			
Waste prevention: • Use modular materials. • Select products for durability. • Eliminate unnecessary finishes and other products.	**Facility:** • Possible decrease in first costs of construction due to less use of materials • Longer lifetime of building and building features due to increased durability of materials used • Decreased operating costs (including replacement and waste disposal costs) by using more durable materials that need to be replaced less often	**Society:** • Less nuisance associated with landfills	**Society:** • Reduced strain on landfills • Reduced use of virgin natural resources
Local/regional materials: • Obtain materials, whenever possible, from local resources and manufacturers.	**Local economy:** • Success of local businesses	**Local economy:** • Employment opportunities	**Society:** • Lower energy consumption and resulting air pollutant air emissions due to less freight transportation
Rapidly renewable materials: • Use materials such as bamboo, cork, wheat straw boards, etc. • Use certified wood.	**Facility:** • Possible reduction of first costs **Society:** • Emerging businesses	**Society:** • Preservation of forests for future generations	**Society:** • Reduced use/depletion of long-cycle renewable materials; better forest management; biodiversity
Design for re-use: • Design buildings for flexible uses so they can be reconfigured in the future as needs change.	**Facility:** • Lower lifecycle capital construction requirements • Lower churn costs	**Society:** • Less disturbance due to new construction	**Society:** • Lower resource consumption (on a lifecycle basis)

SUSTAINABLE DESIGN FEATURE	ECONOMIC BENEFIT	SOCIAL BENEFIT	ENVIRONMENTAL BENEFIT
Indoor Environmental Quality			
Indoor air quality: • Control pollutant sources (from neighboring buildings, soil such as radon and methane, and excessive dampness). • Use low-emission materials (paint, carpet, fabrics). • Allow new materials to ventilate before occupancy. • Include good ventilation systems and operable windows, where appropriate. • Specify systems that feature economizer cycles. • Consider underfloor air ducting. • Use CO_2 sensors. • Use control systems, including personal workstation control, if possible. • Eliminate tobacco smoke.	**Facility:** • Organizational productivity improvements (reduced absenteeism) • Lower worker's compensation, disability, health, and liability insurance costs • Reduced threat of litigation **Employees:** • Lower health care costs to occupants	**Employees:** • Reduced adverse health impacts (especially respiratory disease/discomfort) • Improved personal productivity	**Employees:** • Better air quality inside the facility **Society:** • Reduced volatile organic compound emissions to the atmosphere due to use of low-emission materials
Good visual quality: • Appropriately use windows, skylights, shading devices, and light shelves. • Avoid glare. • Combine ambient and task lighting. • Use high-frequency ballasts. • Give occupants adequate visual access to outdoors and to the organization of the building. • Use pleasing surface colors and reflectivity.	**Facility:** • Reduced energy costs (by using daylighting)	**Employees:** • Satisfaction with work environment • Improved personal productivity	**Society:** • Lower energy use and emissions due to use of daylighting and energy efficient lighting

SUSTAINABLE DESIGN FEATURE	ECONOMIC BENEFIT	SOCIAL BENEFIT	ENVIRONMENTAL BENEFIT
Indoor Environmental Quality, Continued			
Noise control: • Reduce noise at the source during design phase (e.g., through orientation, building layout, selection of mechanical, plumbing, and ductwork). • Use acoustic buffers; floating floor slabs; and sound-insulated wall, floor, and ceiling penetrations. • Achieve good room acoustics by configuring rooms, using white noise, etc.	**Facility:** • Lower cost of dealing with complaints	**Employees:** • Satisfaction with work environment • Improved personal productivity	
System controls: • Provide maintenance staff and users with level of control over automated building systems appropriate to their level of technical expertise. • Provide individual occupant controls when possible (in conjunction with underfloor systems).	**Facility:** • Possible decrease in operational costs • Decreased churn costs due to underfloor systems	**Employees:** • Thermal and visual comfort of occupants • Improved personal productivity	**Society:** • Decreased energy use and resulting environmental impacts due to better control of energy systems

SUSTAINABLE DESIGN FEATURE	ECONOMIC BENEFIT	SOCIAL BENEFIT	ENVIRONMENTAL BENEFIT
Commissioning and O&M			
Commissioning and O&M: • Use third-party assessments to ensure that the installed systems work as designed. • Develop O&M manuals and train staff	**Facility:** • Fewer equipment breakdowns and downtime costs • Lower lifecycle replacement costs • Reduced costs of dealing with occupant complaints	**Employees:** • Occupant satisfaction with building • Health/safety of building occupants	**Society:** • Lower energy consumption and air pollutant emissions **Employees:** • Better indoor air quality
Sustainable housekeeping and maintenance: • Clean and maintain all building equipment to ensure proper functioning. • Check for water leaks and make repairs. • Check for signs of mildew and mold growth. • Use nontoxic, natural cleaning/ maintenance chemicals. • Periodically or continuously monitor indoor air quality.	**Facility:** • Possibly lower operating expenses (e.g., reduced water consumption) • Less chance of sick building syndrome and associated legal costs	**Employees:** • Improved indoor environmental quality • Better health of occupants	**Society:** • Reduced chemical influx into the environment (by using natural cleaning products) • Lower water use (due to leak monitoring)

Additional Resources

This module presents thorough resources for task training. The following resource material is suggested for further study.

California Stormwater Best Management Practices Construction Handbook/Portal. California Stormwater Quality Association, 2009. Available at: www.casqa.org.

Contractors Guide: Save Money and Resources Through Job-Site Recycling and Waste Prevention. Seattle/King County, WA, 2002. Available at: http://your.kingcounty.gov.

Environmental Resource Guide. American Institute of Architects (AIA). New York, NY: John Wiley & Sons, 1996.

Field Guide for Sustainable Construction. Department of Defense – Pentagon Renovation and Construction Office, 2004. PDF, 2.6 MB, 312 pgs. Available for download at http://renovation.pentagon.mil.

Green Building Handbook: A Guide to Building Products and Their Impact on the Environment. Tom Woolley, Sam Kimmins, Paul Harrison, and Rob Harrison. London, UK: E & FN Spon, 1997.

Green Building Materials: A Guide to Product Selection and Specification. 2nd ed. Ross Spiegel and Dru Meadows. New York, NY: John Wiley & Sons, Inc., 2006.

Green Building Product Certification & Labeling Systems. Green Building Alliance, 2009. Available at: www.pa-greenbuildingproducts.org.

Green Construction: Introducing Green Buildings & LEED to Contractors. BuildSmart – Greater Vancouver Regional District, 2004. Contains examples of completed LEED worksheets for construction-related credits. Available at: www.infrastructure.alberta.ca.

Greening Federal Facilities, 2nd ed. U.S. Department of Energy Federal Energy Management Program, 2001. PDF, 2.1 MB, 211 pgs. Available for download at www.ofee.gov.

GreenSpec Directory: Product Directory with Guideline Specifications. 4th ed. Larry Strain. Alex Wilson & Nadav Malin (eds.). Battleboro, VT: BuildingGreen, Inc., 2003.

Handbook of Sustainable Building: An Environmental Preference Method for Selection of Materials for Use in Construction and Refurbishment. D. Anink, C. Boonstra, and J. Mak. London, UK: James & James Ltd., 1996.

IAQ Guidelines for Occupied Buildings under Construction. 2nd ed. Sheet Metal and Air Conditioning Contractors' National Association (SMACNA), 2007. Available at: www.smacna.org.

Innovation in Design Credit Catalog. Washington, D.C.: U.S. Green Building Council, 2009. Available at: www.usgbc.org.

King County Green Tools: Construction & Demolition Recycling. King County, WA, 2010. Available at: http://your.kingcounty.gov.

LEED Reference Guide for Green Building Design and Construction. Washington, D.C.: U.S. Green Building Council, 2009. Available at: www.usgbc.org.

Natural Capitalism. Lovins, A., Hawkin, P., and Lovins, L.H. Boston, MA: Little, Brown, & Company, 1995. Available online at www.natcap.org.

Sustainable Buildings Technical Manual. Public Technologies, Inc./U.S. Department of Energy, 2006. PDF, 3.1 MB, 292 pgs. Available for download at www.smartcommunities.ncat.org.

Sustainable Landscape Construction: A Guide to Green Building Outdoors. 2nd ed. J. William Thompson and Kim Sorvig. Washington, D.C.: Island Press, 2007.

The Green Guide to Specification. Jane Anderson, David Shiers, and Kristian Steele. Watford, Hertfordshire, UK: BRE Press, 2009.

The Seven Sins of Greenwashing: Environmental Claims in Consumer Markets. TerraChoice Environmental Marketing, Inc., 2009. Available at: http://sinsofgreenwashing.org.

Figure Credits

Topaz Publications, Inc., Module opener

© 2010 Photos.com, a division of Getty Images. All rights reserved., 201F01, 201F04, 201F05, 201F08, 201F10, 201F12, 201F13, 201F15, 201F26 (photo), 201F32, 201F36, 201F39, 201F62, 201F67, 201F70–201F74, 201F76, 201F77, 201F79, 201F81, 201SA01–201SA12, 201SA14–201SA19

© Dake/Wikimedia Commons, 201F02

Annie Pearce, 201F09, 201F11, 201F14, 201F20, 201F28, 201F48, 201F56, 201F63, 201F66, 201F68, 201F75, 20178, 201F80, 201F82, 201F87, 201T01-201T11

Sushil Shenoy/Annie Pearce, 201F24, 201F35, 201F57, 201F59

Chris Fiori, 201F06, 201F07, 201F38, 201F40–201F42, 201F44, 201F45, 201F46 (light fixture), 201F47, 201F50-201F55, 201F58, 201F60, 201F61

Sushil Shenoy, 201F17, 201F22, 201F23, 201F25, 201F29, 201F31, 201F33, 201F37, 201F49, 201F69, 201SA13

Art Wickman, CIH, 201F34

© U.S. Green Building Council, 201F18, Appendix A

Image provided by Tate Access Floors, Inc., 201F30

Green Seal, 201F43

U.S. Environmental Protection Agency, 201F46 (ENERGY STAR logos)

The Haskell Company, 201F64, 201F65, 201F83-201F86

Office of Energy Efficiency and Renewable Energy, U.S. Department of Energy, Appendix B

CONTREN® LEARNING SERIES — USER UPDATE

NCCER makes every effort to keep its textbooks up-to-date and free of technical errors. We appreciate your help in this process. If you find an error, a typographical mistake, or an inaccuracy in NCCER's Contren® materials, please fill out this form (or a photocopy), or complete the online form at www.nccer.org/olf. Be sure to include the exact module number, page number, a detailed description, and your recommended correction. Your input will be brought to the attention of the Authoring Team. Thank you for your assistance.

Instructors – If you have an idea for improving this textbook, or have found that additional materials were necessary to teach this module effectively, please let us know so that we may present your suggestions to the Authoring Team.

NCCER Product Development and Revision

3600 NW 43rd Street, Building G, Gainesville, FL 32606

Fax: 352-334-0932
Email: curriculum@nccer.org
Online: www.nccer.org/olf

☐ Trainee Guide ☐ AIG ☐ Exam ☐ PowerPoints Other ______________________________

Craft / Level: Copyright Date:

Module Number / Title:

Section Number(s):

Description:

Recommended Correction:

Your Name:

Address:

Email: Phone:

Index

T

U

V

W